Veteran Bring Backs Volume II

ISBN: 978-0-9727815-5-8

Published By:

Brad
Simpson
Publishing

(A Division of Simpson Ltd.)
140 South Seminary Street
Galesburg, IL 61401
Phone: 309-342-5800
FAX: 309-342-5730
E-Mail: brad@simpsonltd.com
Web: www.simpsonltd.com

Dust jacket design & layout by Brad Simpson & Nicole Browne.
Items on the cover provided by Eric Good and Simpson Ltd.

Editor for the publisher: Carol Simpson

Veterans' stories have been edited as little as possible. Variations in syntax and writing styles have been preserved to help capture the characters of the authors.

Printed in Thailand by:
Thai Watana Panich Press Co., Ltd.
Web: www.twppress.com

ACORN-10
CT-02
CT-02

Every attempt has been made to be accurate throughout this book, and although I have taken every precaution, and have had tremendous help from fellow collectors, I take full responsibility for any errors or omissions herein.

PHOTOGRAPHIC CREDITS

Photos are courtesy of each individual author; and permission was given prior to publication for all stories and photos.

A few specific acknowledgements: a few collectors went out of their way to share their collections, Todd Gustafson, Mel Ashford, Mark Castel, Rob Rhodes and several others, were very helpful in getting this into written form.

A friend G. Terrell took time from his busy schedule to initially proof read the book for me, and then a thorough proof reading by my wife and life-long companion, Terry Pope-Tinker, and then by Carol Simpson, a friend of mine.

Finally, any mistakes found are my fault and no one else's.

Front cover photo is courtesy of Brad Simpson of Simpson LTD

This is the second volume of *Veterans Bring Backs*. The concern when doing a second volume is that the "new" stories will be uninteresting or you will have to compromise in authentication of stories. In other words, poor provenance. However, I have found that the stories in this volume are as fresh as the first volume, interesting in their own right and in fact I discovered several collectors who specialize in finding bring-backs with solid provenance!

My friends on the Japanese forum, many of whom are members of Banzai, a monthly publication by Lodestar Publications, provided a vast amount of verifiable and provenance filled stories. The forum is hosted by Vic Thomas and Brent Snodgrass of **http://forums.gunboards.com** an excellent resource for collectors. This same forum board hosts Jan C Still's **www.Lugers.gunboards.com** and my many friends on **www.Lugerforum.com** have helped me tremendously.

The stories in these pages are generally in the words of the person who sent me the pictures and text and I have changed them only slightly for readability or slight grammatical usage. These stories come from people who were given the different guns, brought them back, or if lucky, bought them and learned the story from the veterans or veteran's families who brought the gun back from the field. Several of the stories are written by people outside of the United States, and I have found that this gives a very unique style and perspective to the stories.

The focus of this book is my fascination with history. Specifically it started with my fascination with the items brought back, and as a teenager I bought bayonets and other items every chance that I got. You will also find an assortment of pistols, revolvers, and rifles in these pages, all guns that our fathers, uncles, and others brought back with them.

Collected by Edward B. Tinker over a period of several years with stories from the people who ended up with the guns and equipment and who could pass the history along.

My name is Andrew Dunn and my great uncle was Jim Gerritts. He was a Marine during the Korean Conflict, serving with the First Marine Division. He was in the Battle of Chosin Reservoir where the Chinese launched a surprise attack against the X Corps, forcing a two week battle as the forces retreated from the reservoir to Hungnam.
I know that he saw extensive action in Korea. If there was some reason that a Chinese or North Korean soldier would have been carrying it, I would not be surprised if Uncle Jim took it, although I don't know why one of the communist soldiers would be carrying a World War II-era Japanese bayonet.

Uncle Jim loved being a combat Marine. He did not fare as well after the Korean Conflict, though. He was stationed at Camp LeJeune for a while, then transferred to Guantanamo Bay,

Cuba, but just did not enjoy the peacetime Marines as much as he did when he was fighting. He always said that the rush of adrenaline during a pitched battle was something that he always missed. So, when his enlistment was up, he got out and got a job with Greyhound driving buses.

He drove for Greyhound for years and also loved that job. I find it absolutely astounding that, as much as he craved the excitement of fighting as a Marine, he was also one of the kindest, gentlest men that I have ever known. He was very funny and very calm. If you did not know that he had seen some of the most horrific combat in history, you would never have guessed it. He was one of the fortunate veterans of Korea who made the transition from pitched battle to civilian life with no problems.

He died, peacefully, late last year. I'll always remember him as one of my favorite uncles.

Author's notes: I know that many Japanese weapons were used in the Korean conflict by the North Koreans and Chinese, as the Japanese left their equipment after WWII.

Shirley Webster was born and raised in southern Iowa, where he lived in the small town of Winterset when WWII erupted. As a 34 year old husband and father of two he wasn't required to serve in the military, but as he told his wife Marie, he didn't think that "he could live with himself if he didn't do his part to help win the war." And so he left his legal practice and in September of 1944, he entered military service as a Lieutenant j.g. in the United States Navy.

After stateside training, Lieutenant Webster was assigned to the USS Nashville (CL-43) where he served as the communications officer from July 8, 1945 to February 16, 1946. Lieutenant Webster participated in the postwar occupation of Shanghai, China when the USS Nashville entered Shanghai harbor on September 19, 1945. While sailing up the Yangtze River towards Shanghai, Lieutenant Webster observed acres and acres of corn fields which made him think about, and greatly miss his home and family in Iowa. Nearing the city, the banks of the Yangtze were lined with children who turned out to greet their liberators. Not knowing which country would be first to reach their city, each child waved a small hand-made flag, which was broken down into four sections, with each section fashioned in the image of one of the victorious nations; the United States, Great Britain, Nationalist China and the Soviet Union. Lieutenant Webster later secured one of these simple flags as a reminder of that day.

While in Shanghai, Lieutenant Webster was given the order to secure the surrender of a Japanese gunboat whose crew refused to surrender to the local Chinese authorities. Taking along a detachment of U.S. Marines to help expedite the situation, Lieutenant Webster boarded the gunboat when a Japanese sailor armed with a Type-38 rifle came out on deck "making a lot of noise." Lieutenant Webster was forced to "silence" the sailor in order to defuse the situation and after securing the crew's surrender, decided to take the rifle back home to Iowa as a souvenir to hunt deer with.

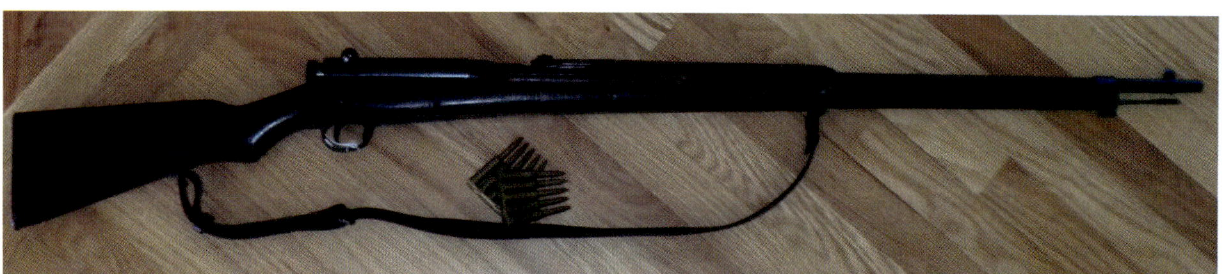

Unground mum on the Type 38

Awards and decorations that Shirley was awarded

Lieutenant Webster and the USS Nashville returned to San Francisco on January 6, 1946. Shortly afterwards, he was discharged from active service and returned home to Iowa. For his military service in the Second World War, Lieutenant Webster was awarded the American Theatre Campaign, Asiatic-Pacific Campaign, U.S. Navy Occupation, U.S. Victory, China Service and the Philippine Liberation Medals.

Other than taking the rifle out to shoot on one single occasion, Mr. Webster kept the rifle in his gun case, perhaps thinking about the day he acquired it or some other distant wartime memory, whenever he glanced upon it while reaching for one of his other rifles or shotguns. Typical of many veterans, he never spoke in detail about how he obtained the rifle as he was obviously bothered by it. To his family, it was a remnant of a time and place that none of them could fully understand or appreciate. As I grew older I expressed an interest in military history and heard that my grandfather, Shirley Webster, had been in WWII. However I never heard mention of the rifle as my family felt it too dangerous or unnecessary for a child of my age to possess. The rifle sat unused year after year until one Christmas when I was around 11 years old, my grandfather told me that he had a special gift for me and he went to go get it. To my surprise, he entered the room with his Type-38 rifle adorned with a huge red ribbon and he handed it to me as my Christmas present.

Now as an adult I have the small flag once waved by a Chinese child on the banks of the Yangtze River welcoming the USS Nashville and its crew to the city of Shanghai, as well as the medals and other mementoes of wartime service brought home by Lieutenant Webster. I have the rifle and ammunition that he also brought back, along with a panorama photo of Shanghai harbor which shows the USS Nashville and another ship marked as "Nashville's prize" - the gunboat that the rifle came from. The rifle itself is a typical early Tokyo arsenal Type-38 complete with Imperial Chrysanthemum, with mismatched bolt and dust cover, that you would probably pass by at a gun show. It isn't a rare or valuable gun, but I wouldn't sell it for a million dollars.

Japanese prisoners

I received this pistol as a gift from the family of the veteran that brought it back. His name was Robert K. LeRouax and he was a First Lieutenant in the U.S. Army. He acquired the gun in December of 1944 during a fire fight in the Battle of the Bulge in the Ardennes. He procured the pistol, holster, and two magazines. He carried it in the holster until he was eventually wounded from a grenade blast. He spent time convalescing in a hospital in Paris, France. Surviving the war he made it home with this nice P38. It is of Spreewerk manufacture built in 1943. The holster has the code EWX over 1943 stamped on the back; EWX was the manufacturers code for Franz U. and Karl Vogels, Holster Manufacturing, located in Cologne Germany.

I am very grateful to have been given this impressive relic.

P.38
201 d

From a small town in Iowa in 1943 to the Atlantic Ocean to join the 215 Maritime Service Training Station at Sheepshead Bay, NY. Yes! That is where I received my shipping papers to go aboard merchant ships as a cook and baker. I was also a butcher by trade and had taken a class in high school in cooking, so I enjoyed the cooking part and still do at age 90.
After my training was over, I was ready to go aboard ship along with some other fellows from the school. Our first ship was a tanker, the Colorado from the Texas Oil fleet.
Our first trip was in a convoy from New York to England. We carried high test gas, P38s, P51s and P47s on our over spar deck. Other convoys to La Havre, France came into play along with one to Bremerhaven, Germany. We discharged our high test gas on one of Hitler's underground gas storage depots. The storage depot was between Bremerhaven and Vegasack and if I can remember, it was called the Wiffo depot. Here I met some US Army soldiers who

were from Graves Interment Station at Vegasack, the ones that came aboard were for the most part from Minnesota. I received a request from one of the big sergeants, asking me if I could find some dry cereals for him and his buddies. He also said that he and the boys would be very grateful if I would fix them up some fried chicken. I got an okay from the Captain to do this for the sarge and his boys for the dry cereals, and being the chief cook, I made sure that they got their fried chicken. I sure did cook a lot of chicken that day.

The next day the sergeant came aboard ship to thank us for the meal, and in return, he gave me a Radom pistol as a token of his appreciate for the meal that I prepared for them. He said that it came from a storehouse in Vegasack, where there were a lot more guns stored there. Yes! I still have my Radom from my trip to Vegasack, Germany in 1945, and with it, the memories of the army guys from Minnesota that wanted some fried chicken. Many times, I often wonder how many of them made it back home safe-n-sound. The Radom will stay in the family as my son has put his bid in already for it and then to his son I hope.

DB Shannon
Merchant Marine Service
1943 -1945

Story brought to the book by Jerry Burney

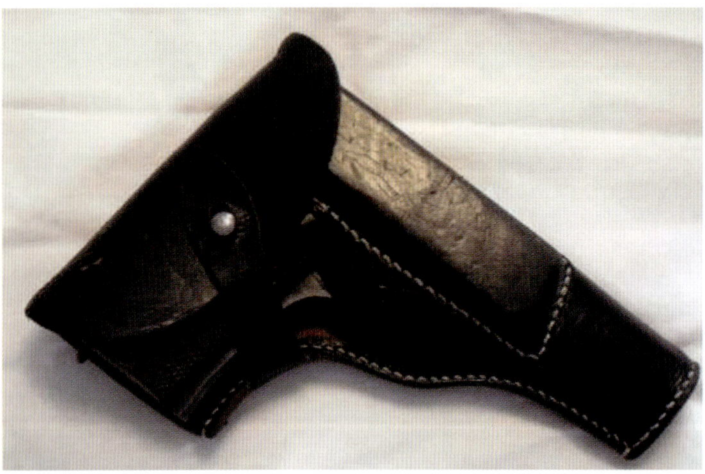

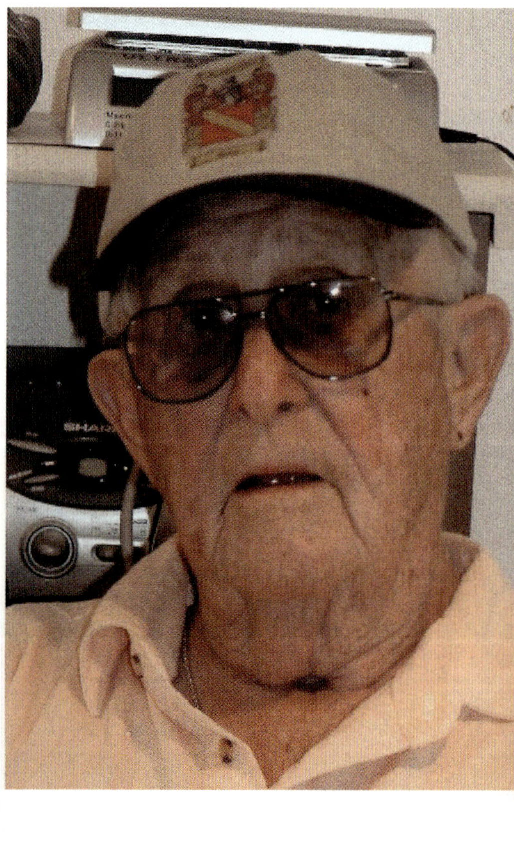

PO/2 Sonar man William E. Marklewitz was the first to make contact with a Japanese hospital ship, the Tachibana Maru in August of 1945, while standing watch on the bridge of the USS Charrette, a Fletcher class destroyer. The ship was boarded in the Banda Sea area. It was seized when the boarding party discovered her cargo marked with red crosses contained hundreds of rifles and small machine guns. Most of the 1663 Japanese soldiers packed into her as "patients" were physically fit for immediate combat.

Only one of the 1663 "patients" couldn't walk from the ship. All others were marched off under guard without incident. They had left New Guinea in small boats for the Kai Islands where the Tachibana Maru took them aboard in the Kai Islands on August 1st to take them to Soerabaja, Java.

William Marklewitz obtained two bayonets and one Type 38 rifle with sling, dust cover and muzzle cap. Apparently he was able to walk by a pile of these items when his ship was in port after escorting the Tachibana Maru.

The rifle is Tokyo arsenal number 1669064 and is matching to include the dust cover, has a leather sling, full mum and brass muzzle cap.

PO/2 Sonar man William E. Marklewitz

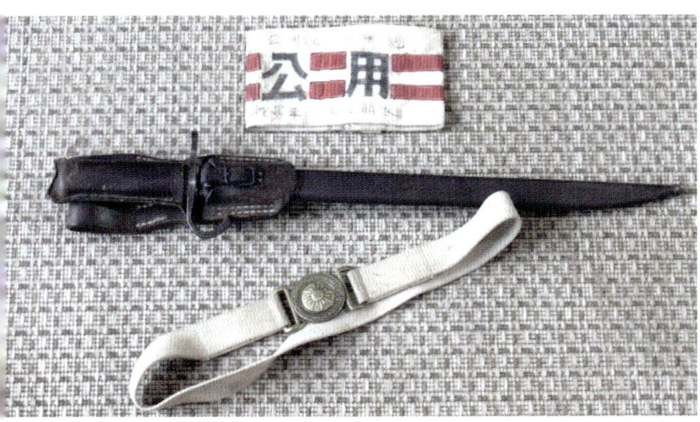

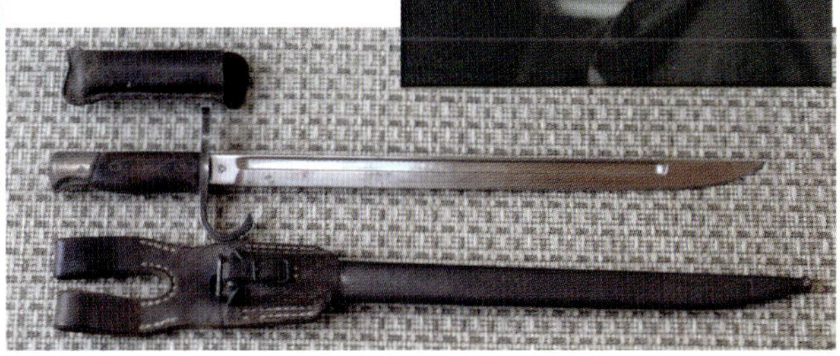

Early in 2005, I acquired an engraved Walther PP from a fellow collector, serial # 215383P. The engraving pattern is the Oak and Acorn type. It came with a shoulder holster. Once I received the pistol and holster I began an inspection of both, and much to my surprise, there was a piece of paper stuffed down inside the holster that said, "Lt. Col. Tedford Blaisdell of Gouldsboro, Me. took this Walther Model P.P. from a dead German officer in Buchenwald."

At that time, through my work, I had access to a large database of people in the U.S., and I decided to see if I could locate Mr. Blaisdell. Since his name was very unusual, and he was located in the state of Maine, I thought he would be pretty easy to locate, and much to my surprise I did locate him! Now it was time to try the phone number I had found for him. Since he was a WWII veteran, I knew it was possible he was deceased, and the little note I found could be the only evidence that could support this story. I dialed the phone, and a woman answered. I asked for Mr. Blaisdell with some doubt, but she said to hold on a second and she would put him on the phone.

I couldn't believe it, as this was 2005 and 60 years after the end of the war. When Colonel Blaisdell answered the phone, I was a little tongue-tied for words, as I didn't really know how to begin the conversation. So, I just went for it, and told him that I was in possession of the engraved Walther P.P., that he used to own.
What followed next was a 20 minute conversation with Mr. Blaisdell that was not only fascinating, but truly a historical account of the infamous Buchenwald concentration camp, and what was found there after the Allied liberation.

I wish I had recorded the conversation, but of course no ones thinks about these things until after it is over and gone. But I did send an email to my friend who sold me the pistol about the conversation, and he was as awestruck as I was.

Following is a summation of what Colonel Blaisdell told me over the phone:
Colonel Blaisdell, as I referred to him over the phone, told me about his acquiring the pistol, and what he saw at Buchenwald. He was in the Army Air Transport in 1944-1945, located in Paris, and he flew out of Orly airport. One day orders came from HQ about flying two U.S. Congressmen to Buchenwald to review the site after it had been liberated by the Allied forces. Buchenwald had been sealed off since the liberation, and the congressmen wanted to review what was there, I guess for both political and marketing purposes. The Colonel was a Major at the time, and could have ordered another pilot to take the congressmen there, but decided to fly this one himself. He mentioned that he had flown Eisenhower on several occasions.

When they got to Buchenwald, he and the Congressmen were touring the camp, and he saw a dead SS Officer leaning against the fence in one part of the camp. He walked up to the body, turned it over, and exposed the engraved P.P. in a shoulder holster still on the body. He took it off the body, and kept it until just recently he said. I asked the name of the Officer, and he just chuckled and said he didn't remember or didn't know. I am sure he didn't.

He said he remembered seeing the lampshades of human skin that the commandant's wife had in their quarters. He also described the piled up dead bodies they saw that were lined up 4-5 high and about 30-40 feet long. He described it as looking like split wood stacked for later burning. When he left Buchenwald, he had the gun/holster, and eventually took it home with him. He said he originally sold the gun to a dealer in Belfast, Maine named Mr. Merry. He sold it after Mr. Merry came around several times asking about it. Not sure how Mr. Merry knew about it, as it didn't come up in the conversation, however after much badgering he sold it to Mr. Merry. It was then sold to my friend, who sold it to me.

At the end of the conversation with Colonel Blaisdell, I asked him if he would send me a letter about the gun. Very much to my surprise, he did, and the letter was received by me around February 16, 2005. In his letter to me, he again stated the serial number of the pistol, and spoke briefly about Buchenwald and horrors of war and what other humans could do to one another.

I sent a letter to the Colonel along with current photos of the pistol the day after I talked to him. About three weeks later, I received a letter from the Colonels' caregiver, Wanda Nichols, who in her letter informed me that the Colonel had passed away just two weeks after he wrote his letter to me. Actually, the Colonel had suffered a heart attack the day he wrote my letter, and lived on and off machines for the next two weeks before succumbing to the heart attack.

In her letter, Ms. Nichols said that the Colonel was very happy that his pistol was in good hands, and he thoroughly enjoyed seeing the photos I had sent of his former pistol, and that it had made his day when I called and talked to him about it. She included his obituary from the local paper.

I have the pistol, the letter from the Colonel, and the letter from his caregiver, and they will occupy a special place in my collection for the rest of my life. It was a most fascinating conversation with the Colonel, and I can only imagine what he saw in Buchenwald to this day. I wish I could have followed up with another conversation or two with the Colonel, but it wasn't meant to be.

TEDFORD M. BLAISDELL, LT. COL. USAF, RETIRED

WEST GOULDSBORO - Lt. Col. Tedford M. "Ted" Blaisdell, husband of the late Margaret (Ashe) Blaisdell, died Tuesday, Feb. 22, 2005, at a Bangor hospital. He was born Feb. 19, 1918, in Newport, the son of C. Carroll Blaisdell and Edith (Bragdon) Blaisdell.

When he was one year old he and his family moved to Franklin. Ted attended Franklin

March 4,2005

Richard Hunter
1100 Rafington Drive
Lawrenceville,Georgia,30045

Dear Mr.Hunter,

I'm writing to you to inform you that Lt. Colonel Tedford M. Blaisdell has passed away.I wanted you to know because it all seemed so interesting how things fell into place with this gun and you calling.I started working for Ted in April of 2004,he of course had many amazing stories.One of them being the story of your gun.I had heard a couple of different times and at the end of it he would always say " I wish I had kept it".So for you to call was very exiting just a good peace of mind for him to know someone had it that really cared for the gun and mostly someone that wanted the history about it.

I want you to know this man was very with it and was living a very full and active life.Well as active as a 86 year old could.He had been putting different paper work off for a few days he just didn't feel like doing it.So the morning of February 12 he got out the paper and started writing to you which I thought was funny because had more important paper work to do than this letter.That was the only paper work he did that day.He put it in the mail within an hour of writing it.Their was something he subconsciously knew he wanted his story told. That was his final words.

That evening at ten o'clock he had a massive heart attack.He lived for ten more days on and off machines.He had his mind up till the last hour of his death,and I am glad to say I was with him when he passed.On the twenty first of February I found your letter in the mail so I took it to the hospital and asked Ted if he would like me to read it to him.He was unable to speak but nodded with a big yes.I read to him your letter it seemed to please him, but when I told him their was pictures he opened his eyes and he gave me a smile I will never forget. He was so happy to see his old gun again.I believe that gun had more of a meaning to him more than just what was in his story,but we will never know.Ted passed away that night at one in the morning.

I have enclosed a copy of his obituary for you, and mostly I want to thank you for taking the time to send the letter and the pictures.You made a dear freind of mine very happy.He was a amazing man.I wish you had the pleasure of meeting him.Take care and enjoy the gun.

Sincerely,

Wanda Nichols

Wanda Nichols

This was in May the 2nd, 1945 in Ludwigslust, Germany, as told by his father LT Col Hampert, by Dan Humpert. While riding with one of my soldiers; Walberg and his squad never got into the fight directly, but they were among the lead American forces across the Elbe river and into southeastern Germany in the spring of 1945. The war was by now coming to a close, with the Allies pushing from the west and the Russians could not be allowed to capture all of Germany. Lead elements of the 82nd Airborne, including their leader General Jim Gavin and Walberg's squad (by then doing reconnaissance) met with the front edge of the Russian army on May 1st, 1945.

Walberg and his jeep driver Paul Schlupp were sent north to scout a highway and determine where the current lines were located. On their trip, they came across a German village of about 10,000 people named Ludwigslust; home to the magnificent palace of the Grand Duke. As Walberg and Schlupp emerged to the north of the village, they began to smell a foul stench. The odor was so bad… We didn't know it at the time, but what we smelled was the odor of the Wobelein Concentration Camp. I told Paul to follow the smell and we came upon a fenced-in camp about four miles north of the village. Like most such camps, the fence was actually two barbed wire fences around a large compound of barracks, housing sick and starving Jews, homosexuals, Gypsies and political prisoners forced to work as slave laborers by the Germans.

When we got to the gate, I could see an emaciated prisoner in ragged uniform laying on the ground, but no sign of any Germans. I asked him in German what this camp was and where the Germans had gone. He told me he was a Dutch political prisoner and that the day before, the German guards left out of the back gate, relocking it and leaving the prisoners alone. We used our jeep to force the gate open and went in to inspect the camp before reporting it to our command. Paul and I stayed there for the next two days and others from our division joined us in assisting these prisoners.

Over 200 bodies were found in the camp according to official records. After two days, Walberg and Schlupp again headed to the north and eventually came within 33 miles of Berlin before they turned back and headed once again for their division headquarters in Ludwigslust. Two days after V-E Day, the 82nd Airborne Division forced the citizens of Lugwislust to dig up the park in the town center to provide a proper burial for the 200 dead found in the camp, plus 800 more that had been buried in shallow graves in the forest adjoining the camp. The adult residents of the village were then required to walk amongst the graves prior to final burial to witness the results of the horror that went on just outside of their town and then the next day, these same adults were paraded through the squalid camp to see what it looked like.

While an MD with the 67th Field Hospital, I acquired a luger that I kept to remind me of the things I had seen.

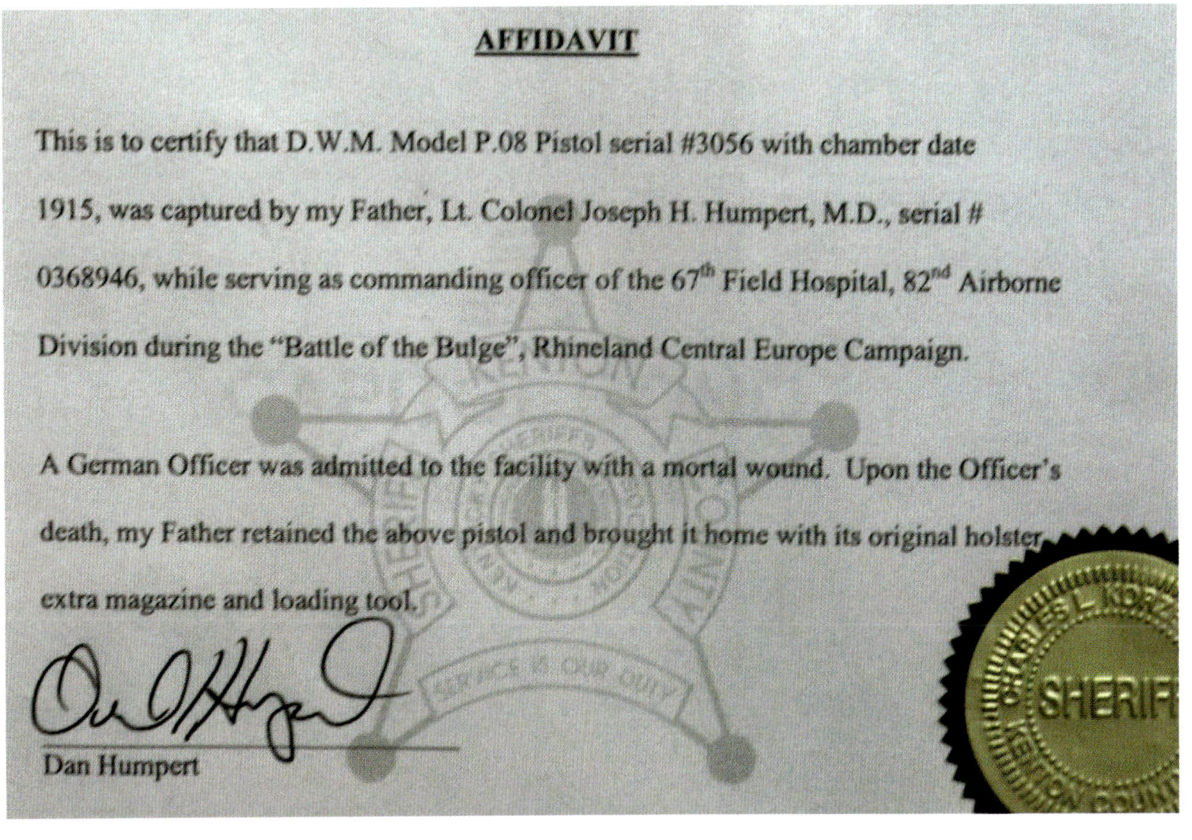

AFFIDAVIT

This is to certify that D.W.M. Model P.08 Pistol serial #3056 with chamber date 1915, was captured by my Father, Lt. Colonel Joseph H. Humpert, M.D., serial # 0368946, while serving as commanding officer of the 67th Field Hospital, 82nd Airborne Division during the "Battle of the Bulge", Rhineland Central Europe Campaign.

A German Officer was admitted to the facility with a mortal wound. Upon the Officer's death, my Father retained the above pistol and brought it home with its original holster, extra magazine and loading tool.

Dan Humpert

Sheriff Chuck Korzenborn-1915 Luger

Photos provided by Capt. Todd Massey, Kenton County Sheriff's Office

US GI with a 1911A1 in the Pacific.

This Czechoslovakian sniper rifle was taken as a war trophy in December of 1969 as a result of a contact between members of the MAT-35 (Mobile Advisory Team - 35) and the NVA. MAT 35 based out of Chu Phu, in the IV Corp area of Vietnam, and operated along the Vietnam Cambodian border. MAT-35 was on the return leg of a 5 day operation into Cambodia, and were observing a NVA supply depot along the Mekong river. After completion of this observation, the team was moving just before sunrise when they came to a clearing. As the team slowed, the point man was taken under fire by a single shot rifle, then by AK-47's. The fire was observed to be coming from a single, lightly fortified bunker. Cover fire was placed on the bunker to recover the point man, then the team began a flanking maneuver on the bunker. Upon completion of the flanking maneuver, the bunker was taken under fire by an M-79 LAW and destroyed. This sniper rifle was recovered from the bunker by the 1st Lt commanding the team and returned to the US as a trophy.

I purchased this rifle directly from the veteran when he brought it into a local gun store to sell. The store owner called me and I went in to look at it. I had never seen a rifle of this type before, and I asked the Veteran if he had any capture papers. He said that he would check, and came back the next day with the capture papers, as well as a box full of photos, his uniform, and the written story of the capture.

The son told me his stepdad had an emergency landing after a bombing raid in Germany. When he landed in a field in France a group of German soldiers approached the downed aircraft. The American GIs were prepared to battle the well armed Germans but instead they surrendered. My best guess is they surrendered to the bomber crew instead of risking getting shot by approaching an American encampment?

This Type 14 Nambu was brought back by Colonel Wendell W. Fertig from the Philippines, along with two copies of the capture papers in his name and displaying the matching serial number. Fertig was cut off from the evacuation of the Philippines, and refused to surrender. He became the commanding officer of the 10th Military District, better known as the Mindanao Guerilla Force, and was continually behind the Japanese lines from early 1942 until April 21, 1945. The capture papers only list the Type 14, although he brought back many barong and kris knives among other souvenirs. I have a ton of other documentation, including 201 file, photos, papers, uniforms, all of which I got direct from his son-in-law.

Although you can't see it, the Nambu has a crudely repaired firing pin; a common problem on the Type 14s, and I suspect this is a guerilla repair job. The recoil springs are also weak. The 12.3 date indicates manufacture in March, 1937. Its finish was a uniform blue that has turned almost completely brown with fine freckling all over, not surprising for a humid climate near salt water.

The Nambu could easily have been (and probably was) actively used by the guerillas after it was captured right up to the end of the war and then presented to Fertig. The guerillas did a lot of hit-and-run attacks on Japanese patrols, grabbing their gear when possible, and killed about 7,000 of them during the occupation. There were only a couple of formal attacks on Japanese strongpoints. I believe the Japanese had about 60,000 troops on Mindanao trying to root out the guerillas, which numbered about 30,000, or so it is said. The plaque they presented to Fertig on July 4, 1945 says 33,000, but I suspect that number includes relatively few full-timers and a lot of sympathizers and hangers-on. Back in the 60s a book was written about Fertig's operation, based on his papers titled "*They Fought Alone*," and a fairly decent movie was also filmed.

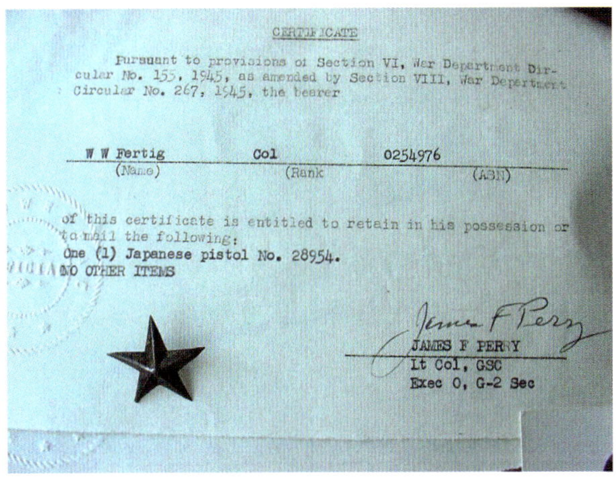

CERTIFICATE

Pursuant to provisions of Section VI, War Department Circular No. 155, 1945, as amended by Section VIII, War Department Circular No. 267, 1945, the bearer

W W Fertig	Col	0254976
(Name)	(Rank	(ASN)

of this certificate is entitled to retain in his possession or to mail the following:
One (1) Japanese pistol No. 28954.
NO OTHER ITEMS

James F Perry
JAMES F PERRY
Lt Col, GSC
Exec O, G-2 Sec

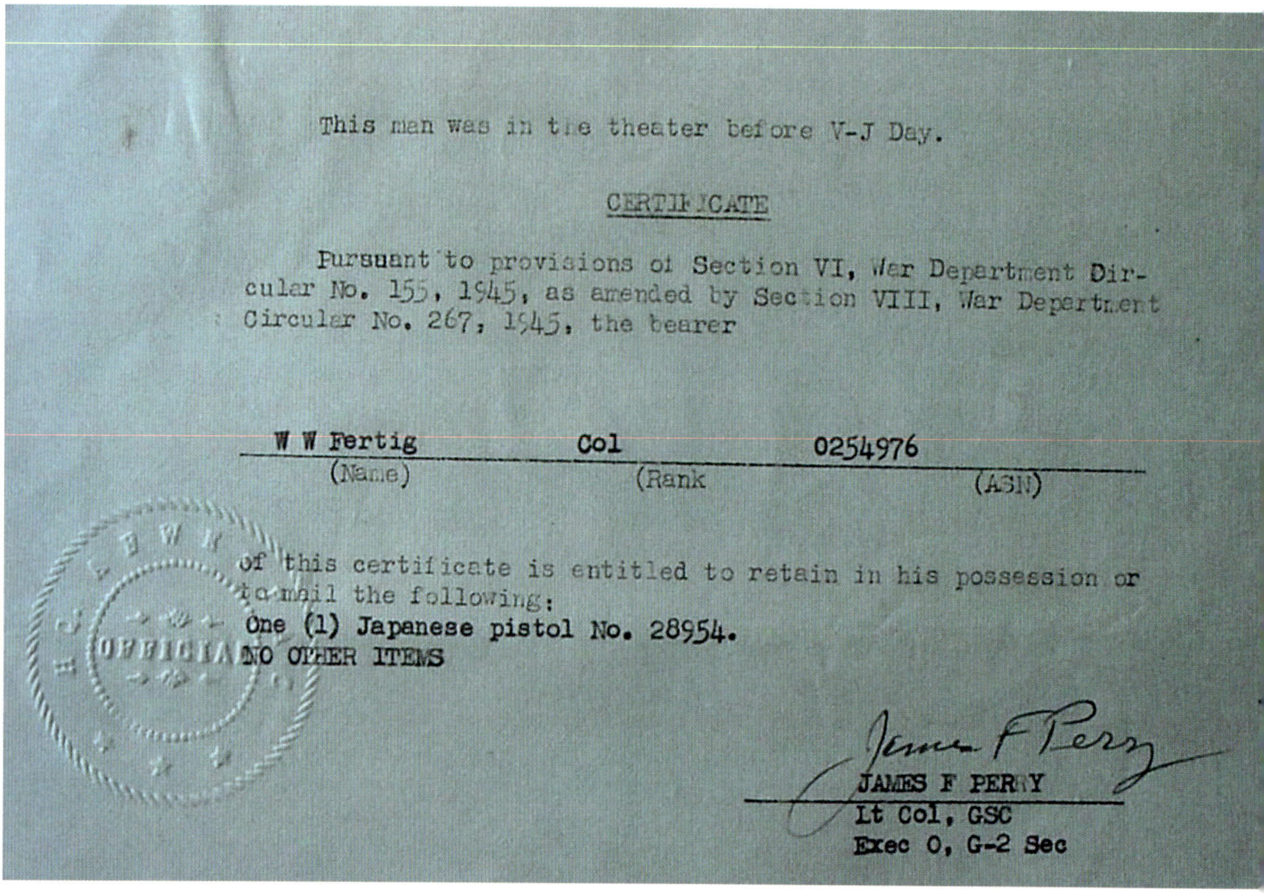

This man was in the theater before V-J Day.

CERTIFICATE

Pursuant to provisions of Section VI, War Department Dircular No. 155, 1945, as amended by Section VIII, War Department Circular No. 267, 1945, the bearer

W W Fertig	Col	0254976
(Name)	(Rank	(ASN)

of this certificate is entitled to retain in his possession or to mail the following:
One (1) Japanese pistol No. 28954.
NO OTHER ITEMS

OFFICIAL

JAMES F PERRY
Lt Col, GSC
Exec O, G-2 Sec

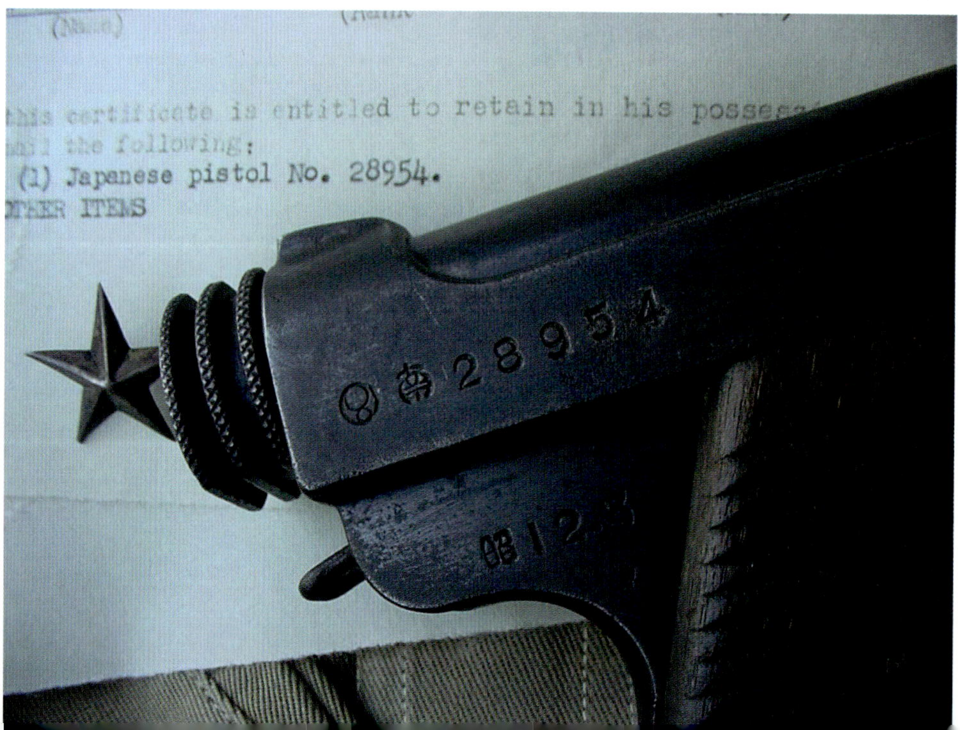

LUGERS FROM NORWAY
****

On April 8, 1992 I purchased three Lugers which had just been imported into Canada from Norway as surplus. They were:

1. "S/42", "G"-Date, serial number 2173f (Collection # 97)
2. "S/42", "1936", serial number 6767k (Collection # 98)
3. "S/42", "1938", serial number 3477i (Collection # 99)

HISTORY:

In early May of 1945 there were still 100,000 Germans in German-occupied Norway, commanded by Lieut. General Franz Bohme, a hardliner who boasted, "In Norway we can accept any battle".

However, on May 7 General Alfred Jodl signed the papers surrendering all German forces on all fronts. One of the conditions was that it required the Germans to destroy no machinery or military equipment.

In spite of the bellicosity of its commander, the German occupation forces in Norway quietly turned over their arms to an Allied expeditionary force only 40,000 strong. General Bohme reconciled his men and himself to surrender with a last order of the day: "Grit your teeth, maintain discipline and good order, obey your superiors and remain what you have always been-German soldiers who love their people and their country more than anything else in the world."

With the surrender of the Germans the Norwegian authorities resumed control of their country. Military and Border Guard units were re-organized and were issued with the surrendered German euipment, including small arms such as the P.08 and P.38 pistols.

The Norwegians kept these pistols as issue, and later as back-up in storage, until 1991 when they were released as surplus.

IMPORTATION INTO CANADA:

These Lugers were imported into Canada by a company called:

Tacair Systems Incorporated,
2367 Anson Drive,
Mississauga, Ontario, Canada L5S 1G1

Tacair (which stands for Tactical Aircraft) is a company which buys and sells fighter planes and other aircraft to governments, companies and private individuals around the world. It is not normally in the gun dealer business.

Its two partners, Barry Johnson and Terry Kingsley (I think its Kingsley) had some kind of aircraft deal with Norway. The Norwegians were short of cash and offered them Norwegian surplus arms as payment. They accepted the deal and took 4,000 Lugers and P.38's because they had at least heard of the P.08 and P.38 pistols. I think they said they took 3,000 Lugers and 1,000 P.38's. They then sold them all to gun dealers in the United States. This was approximately the middle of 1991 and these pistols then flooded onto the American market in the Fall of 1991. Reference was made to them in the October, 1991 issue of "Auto Mag".

Tacair imported another 58 pistols from Norway in March of 1992. This batch consisted entirely of Nazi issue P.08's with German holsters, spare magazines, loading tools and Norwegian cleaning rods. The holsters had been modified by the Norwegian military so they could be carried on the standard Norwegian military belt which had eyelets in it like the U.S. military belt. The original belt loops were carefully removed and then a steel wire assembly was attached using four rivets. The original German markings had not been removed and those which had been placed high enough were still visible on the back.

The Lugers were varied, starting with some "1920" D.W.M.'s of Weimar vintage, through four Krieghoff's and one Simson to Mauser specimens from "G" to 1942. Average condition was Very Good to Excellent, with most of them having all matching numbers. There were a couple of re-blues and a couple of barrel replacements. None of them had any obvious Norwegian ownership markings and none of the original German markings had been removed. A couple of them had small marks like "S (A) E" on the underside of the barrel which I presumed meant re-barrelling by the Norwegians.

The R.C.M.P. yellow dealer cards show that these Lugers were obtained from: The Norwegian Armed Forces Museum,
Oslo Mil/Akershus,
0015 Oslo 1, Norway
on March 19, 1992 and were "N.P.R." = Not Previously Registered.

I first met Barry and Terry at the O.A.C.A. gun show at the Jane St. Bingo Hall south of Hwy. 7, on Sunday, March 22, 1992, where they were displaying some of their Lugers and accessories. Barry Johnson gave me his home phone number and on March 22 I phoned him. He said the Lugers were at Tacair, a small industrial buiding near Toronto International Airport. Maria and I drove there on Sunday, March 29. There was another collector there named Fred Lesage (a NAPCA member) who was helping them classify the Lugers. Tacair preferred to sell all 58 Lugers as a single lot but they were willing to sell them individually. The total retail value of the lot was $40,000 and they wanted to get 25 or 30,000 dollars.

Barry said he can get more stuff from the Norwegians and that he is even thinking of setting up a retail business selling this sort of stuff since the aircraft business doesn't keep him busy enough.

On April 8/92 I visited Tacair again for a closer look at specific Lugers I was interested in. The Krieghoffs were expensive ($1,900) and the Simson and D.W.M.'s weren't in excellent condition so I picked out 3 S/42"s, a "G"-date, a 1936 and a 1938, all in excellent condition. They wanted $1,900 for just the pistols. I countered with an offer of $1,200. They then asked $1,700 and would include a holster, spare mag, loading tool and cleaning rod. I accepted, since my price research showed the current retail value of a 95% P.08 alone to be $682. Also, he agreed to pay half the taxes. P.S.T.=8% plus G.S.T.=7% is 15%. So I paid 7.5% tax.
Total cost was $1,700 + 7.5% = $1,827.50 for the 3 Luger rigs.

On April 9 I paid in cash and registered my new Lugers.

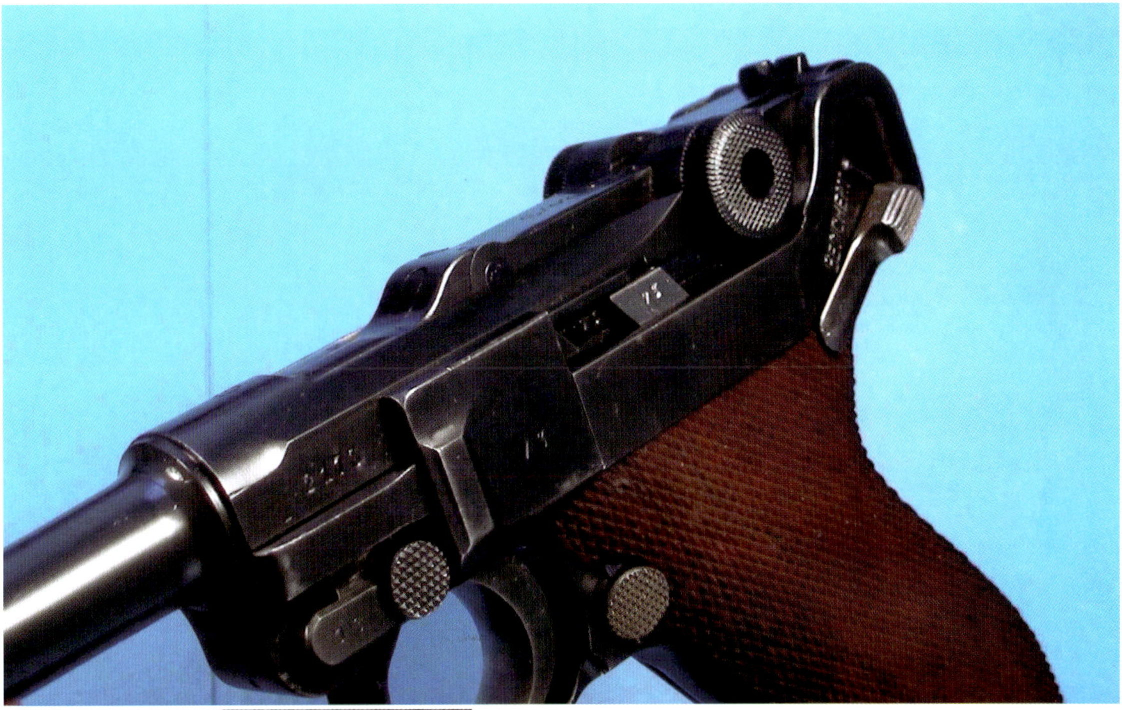

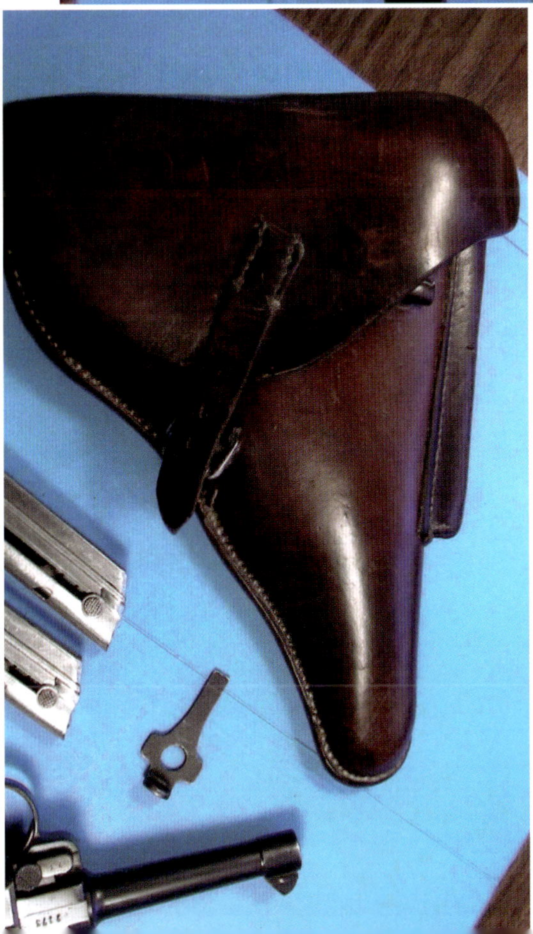

S/42
73

2173

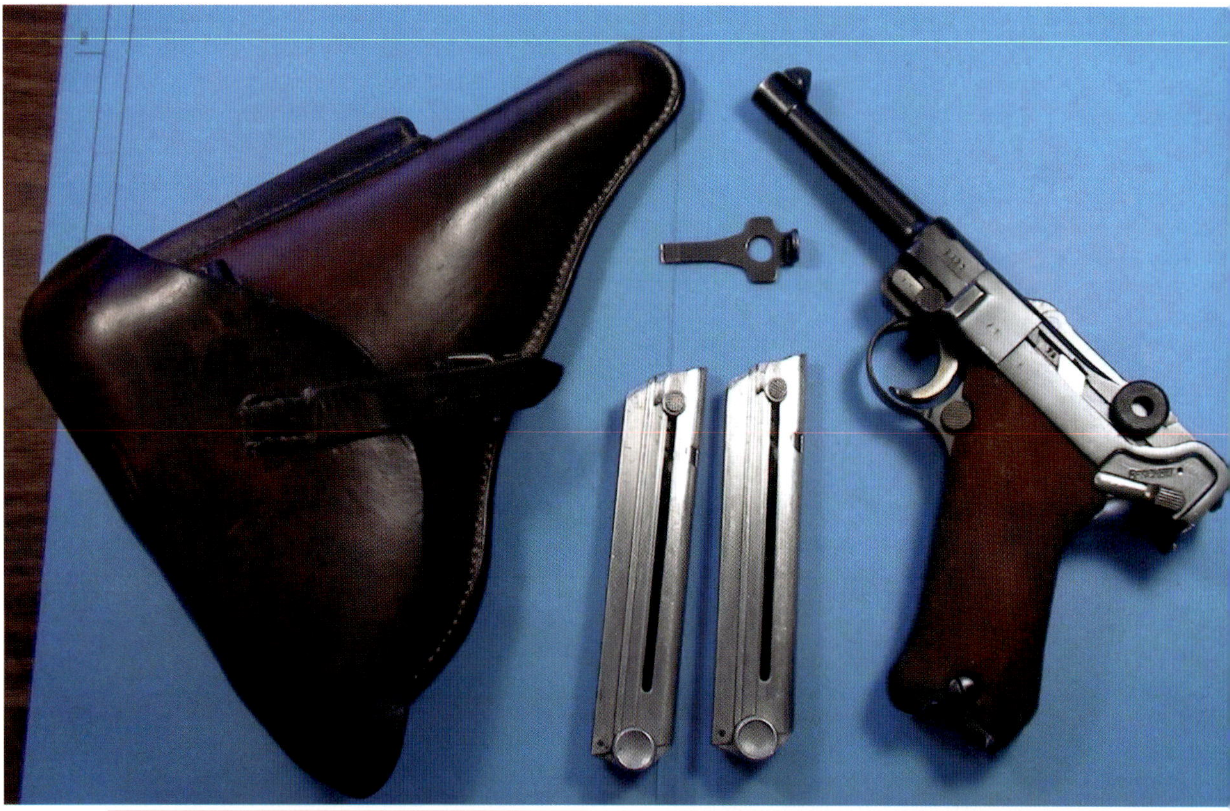

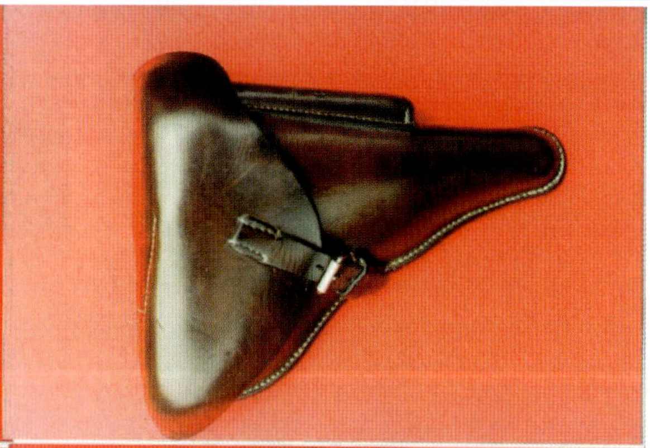

Nowadays, the Brazilian contribution to the Allies' war effort during World War II is practically unknown outside of the academic environment. But the fact is that, in January 1942, Brazil broke relations with the Axis at the Rio conference, and entered the war officially in August of that year, after they had several merchant ships torpedoed and sank by the German U-boats.

Named "Força Expedicionária Brasileira" (Brazilian Expeditionary Force {FEB}) the troops sent to Italy in five echelons eventually totaled 25,334. In July 1944, the first echelon arrived in Naples, on Sept 15, 1945, the 6th Infantry Regiment and support troops, under Brigadier General Euclydes Zenobio da Costa, went into the line of the Fourth Corps of the U.S. Fifth Army, under General Mark Clark.

FEB role was a tactical force; the bulk of its combat experience was at the platoon level. Even in this limited role, the Brazilians performed very well, and many of its soldiers were awarded for gallantry by the US Army: around 180 Bronze Star Medals, 30 Silver Stars and one Distinguished Service Cross were awarded among its ranks.

Among the battles fought by FEB in Italy, reference must be made to "Monte Catello" ("Castle hill"). Monte Castello is a minor elevation lost amidst some of the ruggedest terrain in Italy. It does not show up on large-scale maps of Italy, but to Brazilian soldiers, its conquest, after four attacks during the winter of 1944-45, had great symbolic importance. The fact is that the FEB and the U.S. l0th Mountain Division were effective in the joint operation which drove the Germans off important elevations that allowed the Allied spring offensive to move forward. lf either of the two divisions had failed, that offensive would have been delayed. Among the Brazilian soldiers who were at Monte Castello, was Major Lelio Rebello Miranda, who brought back this Mauser Model 1914 and its holster.

As a general rule, Brazilian soldiers were not allowed to bring back any captured guns from Italy. History shows that this was not always enforced and several pistols made their way to Brazil as part of the FEB soldiers war mementos. Since these 'bring-backs" were undocumented; there is no such a thing as the US "capture papers" linking the guns to any particular story.

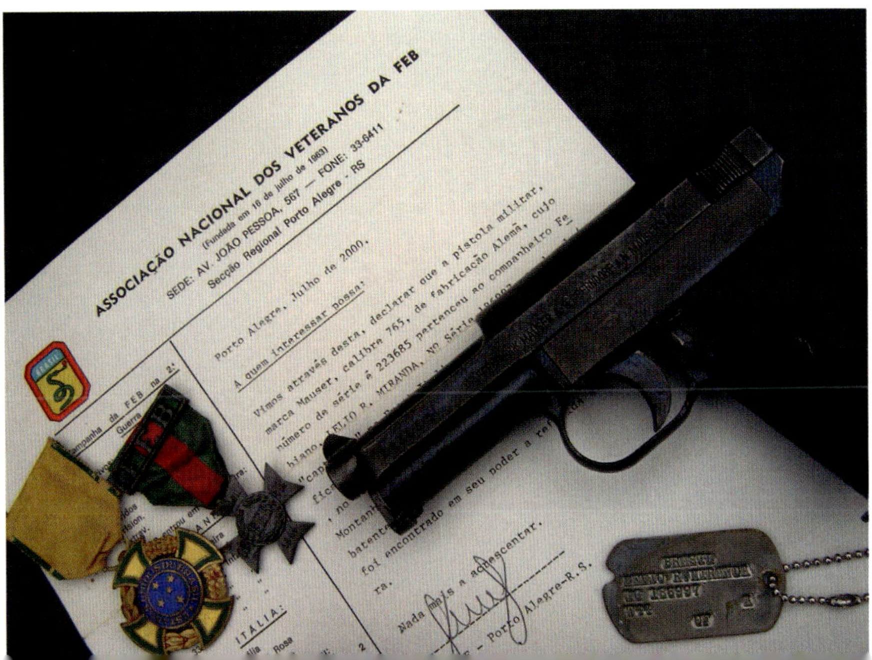

The medals shown are the ones mentioned in the text: Brazilian War Medal (with the yellow ribbon) and FEB Campaign Medal (with the green/red ribbon), both won by Major Miranda.

Notwithstanding, this gun is an exception to the rule. The rig, comprising a Mauser pistol Model 1914 and its brown leather holster, it is has an affidavit from the FEB Veterans National Association where the circumstances of its capture is recorded. According to the document:

"...we declare that the military pistol brand Mauser, caliber 7,65, from German manufacture, which serial number is 223685, owned by FEB comrade LELIO R. MIRANDA, service number 186997, was captured on Italian Front in the year of 1944, nearby Monte Castello, at the advanced medical station of the 10th US Mountain Division, when he, upon the searching of a wounded and fainted German Airforce soldier, found the above mentioned pistol." McCann, Frank D. "Brazil and World War II: The Forgotten Ally. What did you do in the war, Zé Carioca?"(found on the http://www.tau.ac.il/eial/VI_2/mccann.htm website on Sept. 18th, 2010).

In addition to the affidavit, holster and the pistol rig, one of the dog tags of Miranda came together with the gun, confirming its service number. Further research on this man, revealed that he was a cavalry officer, that he shipped to Italy on November 23rd, 1944 and returned to Brazil on September 17th, 1945. Miranda served most of his time in rearguard duties medical hospitals, probably as a liaison officer, which would explain his presence at the 10th US Mount Division post.
Miranda left the Army right after his return to Brazil, and kept the pistol as a token of his service in the Italian front in 1944-45. For such services he was awarded with the Brazilian Campaign Cross and the Brazilian War Medal. Today the whole rig stays in my collection as one of the few documented Brazilian war trophies known to exist in private hands.

In July 1967, I was in South Vietnam doing EOD and Bomb Damage Assessment for the Navy Department. I had extensive training and experience with both nuclear and conventional weapons. This was my third trip in-country. I was at the time an Enlisted Bombardier/Navigator Instructor in the Douglas A3 Skywarrior assigned to VAH-123 at the Whidbey Island Naval Air Station, at Oak Harbor, Washington.

When I was assigned a mission, my EOD team, which consisted of 10 members, would be accompanied by either a squad of Marines or Army personnel for security.

On one such mission we were sent to a location near the DMZ to recover unexploded 500# GP bombs. We would recover the bombs, remove the fuses and detonate the duds to prevent the VC from using the explosives for making booby traps.

I carried an M14 after a very bad experience with an M16 and also a Colt M1911A1 in .45 acp. Although I was technically a non-combatant, the Viet Cong (VC) did not understand that and would try to kill me!

As we were making our way to the site, we ran into a group of VC and a firefight ensued. I tried to keep my men and me out of the way, however as I stepped from behind a large tree, I came face to face with a young VC who had an AK 47 slung over his shoulder and had a revolver in his right hand.

As he raised the revolver in my direction, I raised the M14 and fired two rounds in quick succession. Both rounds hit him in the chest. He fell back against a tree trunk and I ran over and kicked the pistol out of his hand.

After the firefight was resolved, I came back to examine the body and found that the pistol was a Smith & Wesson Victory Model. I removed the remaining unfired rounds and saw that the nose of a bullet was protruding from the muzzle. Also the barrel had lumps along it's length, so I suspected that it had rounds stuck in the barrel. I dropped the S&W into my pack and continued with the mission. I brought the pistol home and using a Dremel Tool, I cut a section out of the barrel and found 9 rounds jammed in the barrel!
I had the Navy photo lab take some photos and mailed them to Smith & Wesson. They were impressed and sent me a nice letter…

I have displayed this pistol for many years at gun shows and it always gets a lot of attention!

John Holbrook ADJC
USN (Retired)

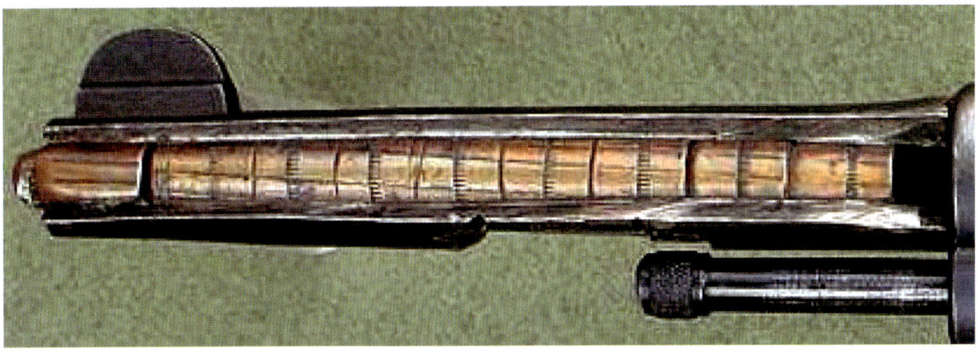

My Dad, Alfred Nielsen, served with the 740th Field Artillery Battalion during WWII. He served with the Headquarters Battery and was the Battery Liaison Agent. I went to the National Archives and copied his unit's history - In fact I copied every piece of paper that the National archives had in their boxes for the 740th - 700 plus (which I have shared with other children whose fathers served in the unit).

He saved many paperwork items (not all) encompassing his service - Starting with his draft notice and ending with the paperwork that he received upon discharge from the Army in Dec. 1945. I have a silk map where he traced his route, and he made a pamphlet of the battle route, and he even wrote down the battle route on foreign money which I still have. I have all 287 letters that he wrote and have transcribed them in word, but I haven't really proof read them as of yet. I have other paper type items as well but they are more minor in nature, but still cool. Among the Nazi items that I have are: Grand Party Badge (#6158 - owner unknown as of now - another project), the Gau Badge for 1923, 3 tinnies, and 8 misc. others, both versions of the Luftwaffe dagger, 1 bayonet, 1 Hitler youth dagger (in rotten condition), 1 Iron cross (I'm pretty sure it's WW1), belt buckle (there may be 2), helmet insignia, Nazi armband, a few other items and the coolest item is a Nazi banner where he had men from his battalion sign their names around the swastika (about 35-45 names). And of course, the Luger. I not only have the Army certificate for the Luger, but I have the certificate for most of the other items listed above as well as him listing these items in a letter that he wrote home on June 8th, 1945.

I have his Army certificate dated November 20, 1945 which lists the Luger as:
"1 German Luger #46"
" S Hn I 2-27 (on Handle)"

The army certificate (AG USFET Form No 33) was needed for him to personally carry the Luger back to the US.

*Editor note: The slightly different markings, *S.Hn.I.2.27.* on the gun and *S.H.I.2.27.* on the spine of the magazine, are typical of the Hannover Schupo Lugers from the early 1920s. The gun was probably marked before the issue of the April 1922 marking instructions specifying the abbreviation for Hannover as H. The magazine is probably a replacement that was marked after 1922. There are a number of similar combinations recorded in Table 9-6 of <u>History Writ in Steel</u> by Don Maus (an excellent book on police markings).*

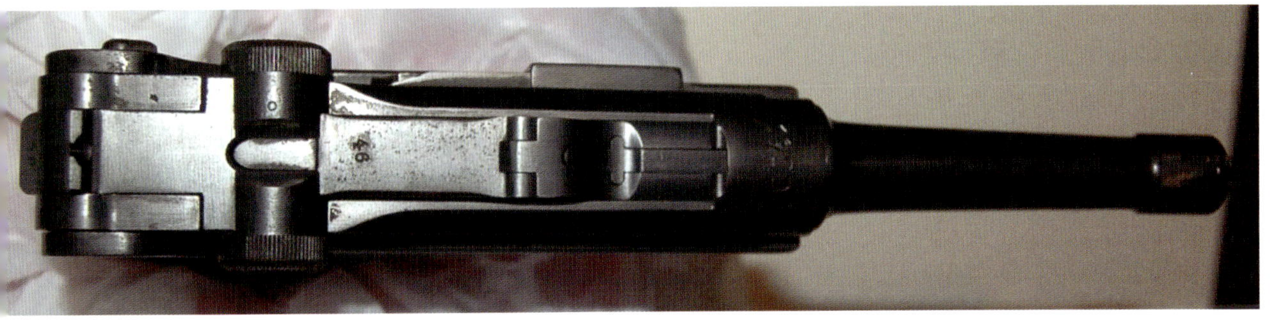

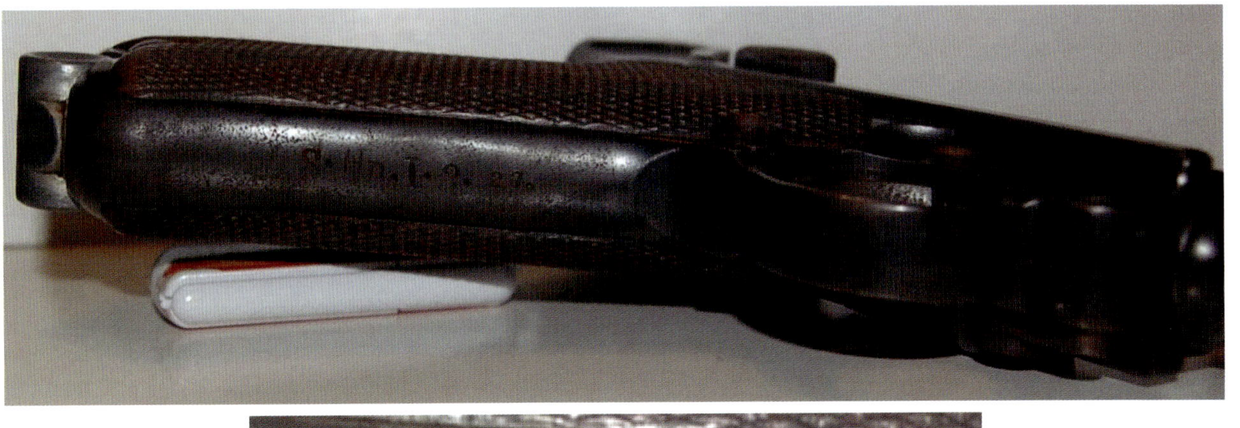

Dad's name was Alfred Nielsen aka Corporal Alfred N.M.I. Nielsen (no middle initial) - Although he was also a Korean war Veteran later attaining the rank of Sergeant (he didn't go overseas this time, he was stationed at the Suffolk County AFB on Long Island, where he met my mom). He had continuous service from 1942 starting with the Connecticut State Guard—Army (WWII) then Army Reserves—Air Force (Korea) and then Air Force Reserves, ending his military service in 1955.

From the Vets friend:
He was a tanker in WWII and said that he took the pistol from a guard on the first day of the siege at Suhl. He was in either the 9th or 11th division. He was in both WWII and Korea.

Almost everybody had these pistols and after the ground pounders established themselves there were non-combat related casualties due to too many people who needed direct supervision with too much time on their hands and a surplus of these pistols and ammo.

Taking a chance, he sent this pistol home with a bunch of toys and was fortunate enough to have honest people in the supply chain in that command to get these things home in one piece. (a lot of things that were to be sent home were stolen).

He said most of these pistols were destroyed by order of the commanding officer after a series of injuries and deaths caused by careless use.

This last ditch Japanese Arisaka concentric circle rifle was brought home from Japan by my mother, 2nd Lt. Alice M. Peterson, who was an Army nurse in the Pacific. She was originally on Tinian to receive casualties from the invasion of Japan but then after the war ended she was part of the occupation force in Japan. She remembers the nights the A-bombs were flown off the island, because the island was blacked out for a long period of time and she has pictures of the Enola Gay undergoing repairs. When my Mom left Japan for the States, the port she left from was not far from Kokura, and on the docks were barrels of rifles and bayonets for souvenirs and she picked up this rifle and a bayonet. All the Marines on the ship were laughing at the sight of all the nurses getting on the ship "armed to the teeth" with Arisakas.

Last ditch concentric circle rifle.

This is the Enola Gay August 1945

Assorted WWII Japanese ammunition that has made its way to the USA

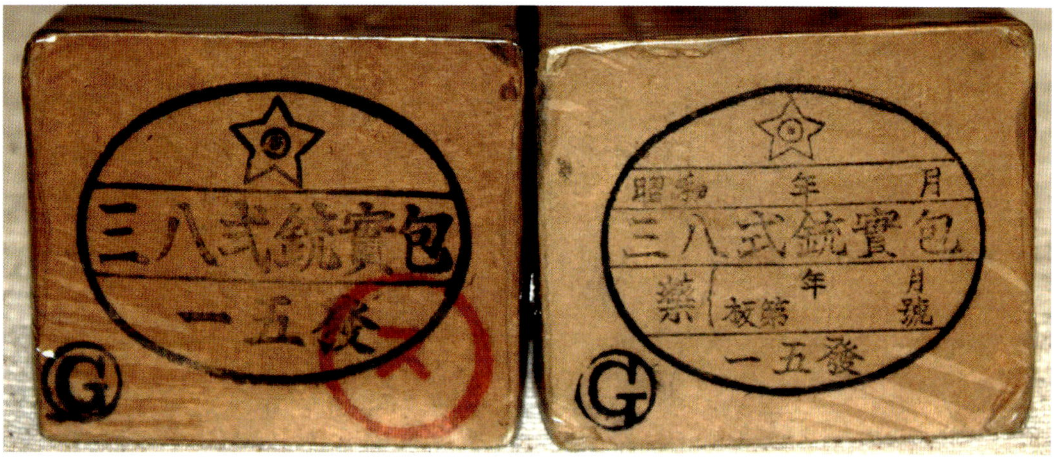

6.5 mm Japanese rifle ammunition

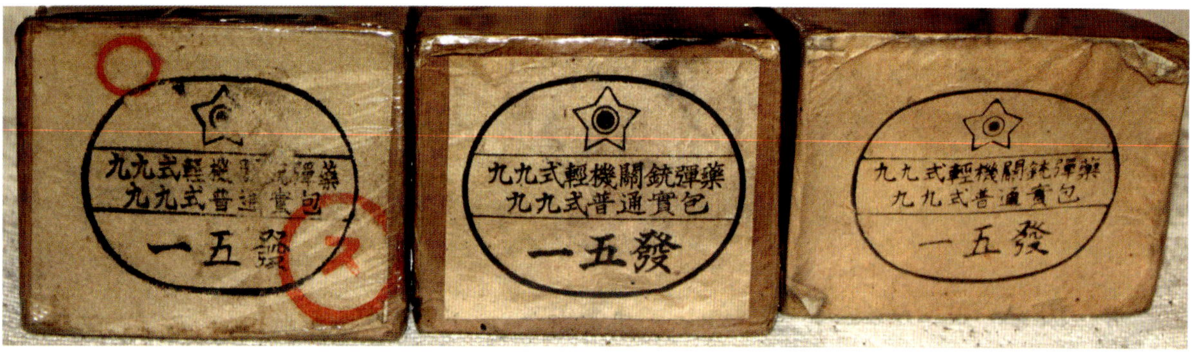

7.7 mm Japanese rifle ammunition

7.7 mm Japanese rifle ammunition that is steel cased (most are brass cased)

This is the story of how I came to own my Dutch contract, FN M1922 pistol serial number 6448 and the GI who brought it home from Europe. I purchased the pistol in May of 2009 because it had the "bring back" documents with it. I was determined to research the pistol and try to obtain as much information on it and the GI as I could.

As for the pistol itself, it's a pretty standard example of a M1922 pistol as produced by Fabrique Nationale in Belgium. The pistol is a Dutch contract pistol, which means it was produced by FN for the Dutch military. In Holland, the pistol was known as "Pistool M25 No.2" and is in 9mm short (.380 ACP) caliber. The pistol has all matching serial numbers including the magazine, and bears the Crown over W crest of Queen Wilhelmina on the top of the slide. Owing to the serial number, this pistol was probably made in 1927 or 1928.

As for the GI, I was fortunate enough to track down some of his information, as well as a picture, and compile this record of his service. Private First Class Benjamin D. Myers Jr. was born on May 19, 1923 in Marion, South Carolina. He was inducted into the U.S. Army on April 30, 1943 in Philadelphia, PA. PFC Myers was sent to Company "D" 1st Battalion, 263rd Infantry Regiment of the 66th Infantry Division. The 66th was known as the "Panther Division" due to the distinctive black panther on their division shoulder patch. PFC Myers had a Military Occupational Specialty of Heavy Machine Gunner. PFC Myers and the 263rd Regiment boarded the Navy transport ship George O Squier and departed New York harbor on the morning of November 15, 1944, arriving in Plymouth, England on November 26th. One month later, the entire 66th Division was in France and ready to get into the war.

The 66th Division was assigned the task of containing the German resistance in the St. Nazaire and Lorient pockets. During the evening of New Years Eve and the day of January 1st, 1945, PFC Myers and the 263rd regiment took over control of the St. Nazaire North sector. It was in this secto that PFC Myers fought until the German surrender on May 8th. By June, the 263rd was operating Camp St. Victoret, a staging and processing camp for troops and equipment heading to the Pacific. By October, the 66th Division had orders to return to the U.S. but, like most men in the 66th, PFC Myers didn't have enough "points" to return home. It was about this time that he was transferred to Company "D" 1st Battalion, 232nd Infantry Regiment of the 42nd Infantry Division for occupation duty in Austria.

In March of 1946, PFC Myers boarded a transport and arrived back in the U.S. on the 3rd of April. Five days later, PFC Myers was honorably discharged from the army at Ft. Dix, NJ and returned to civilian life. During his 3 years and 6 months of service, PFC Myers was awarded the Combat Infantryman Badge, American Campaign Medal, European, African, Middle Eastern Campaign Medal, Army Good Conduct Medal and the World War II Victory Medal.

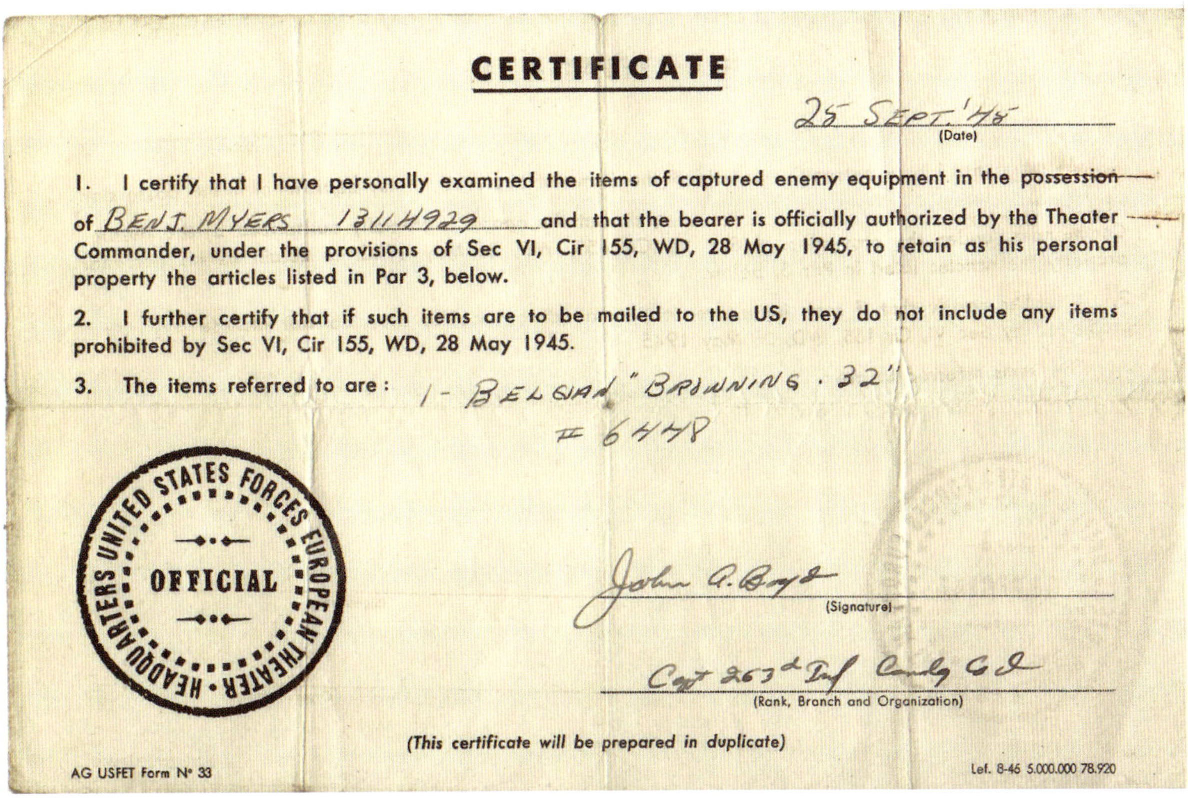

CERTIFICATE

25 Sept. '45
(Date)

1. I certify that I have personally examined the items of captured enemy equipment in the possession of Benj. Myers 13114929 and that the bearer is officially authorized by the Theater Commander, under the provisions of Sec VI, Cir 155, WD, 28 May 1945, to retain as his personal property the articles listed in Par 3, below.

2. I further certify that if such items are to be mailed to the US, they do not include any items prohibited by Sec VI, Cir 155, WD, 28 May 1945.

3. The items referred to are: 1- Belgian "Browning .32" #6448

HEADQUARTERS UNITED STATES FORCES EUROPEAN THEATER
OFFICIAL

John A. Boyd
(Signature)

Capt 263rd Inf Cmdg Co
(Rank, Branch and Organization)

(This certificate will be prepared in duplicate)

AG USFET Form N° 33

Lef. 8-46 5.000.000 78.920

TREASURY DEPARTMENT

Internal Revenue Service

Alcohol Tax Unit

Name:

Address: Benj. D. Myers, Jr.
101 N.E. 4th St. Milford, Dela.

Dear Sir:

A permanent record has been made by this office; under date of 9-26-46 , of the fact that your war souvenir firearm:

DESCRIPTION:
Belgium Semi-auto Pistol, Ser#6448, 9 m/m

has been inspected, and that the same does not require registration under the National Firearms Act.

For information relative to the Firearms Regulations for the State of ~~New Jersey~~ Delaware, it is suggested that you make inquiry at the State Police or Municipal Police Department.

The Alcohol Tax Unit is anxious that you enjoy the possession of your souvenir. However, it is also anxious that you have no misfortune with the firearm. It is suggested that you observe all common sense precautions in the displaying and handling of the firearm, in order to protect yourself and your family from possible serious injury or death by carelessness or misuse.

The Alcohol Tax Unit, of the United States Treasury Department, is charged with the enforcement of the Internal Revenue Code as it relates to illicit alcohol as well as to the National Firearms Act. If at any time information relating to manufacture, possession, or transportation of illicit alcohol reaches your attention, we would appreciate your getting in touch with this office.

Very truly yours,

Michael A. Dessolano
P.C.C.

Investigator in Charge
Rm 426 - P. O. Bldg
Camden, N. J.

(1155)

Here's a Type 38 rifle which I have in my collection. I have been in close contact with "Muddy Gentili", who is still among us (retired cop and a very amazing guy). I have his written story about this rifle, which is not very exciting, but it is a typical story for these sorts of souvenirs. "Muddy" Gentili's rifle was obtained in a trade with some Marines. He was a parachute rigger aboard CV 50 USS Bennington (an aircraft carrier) and has carved all of the places the ship was engaged in on the stock. Muddy told me that the rifle was confiscated in Hawaii in 1945, and he never saw it again until I sent him a picture of it three years ago.

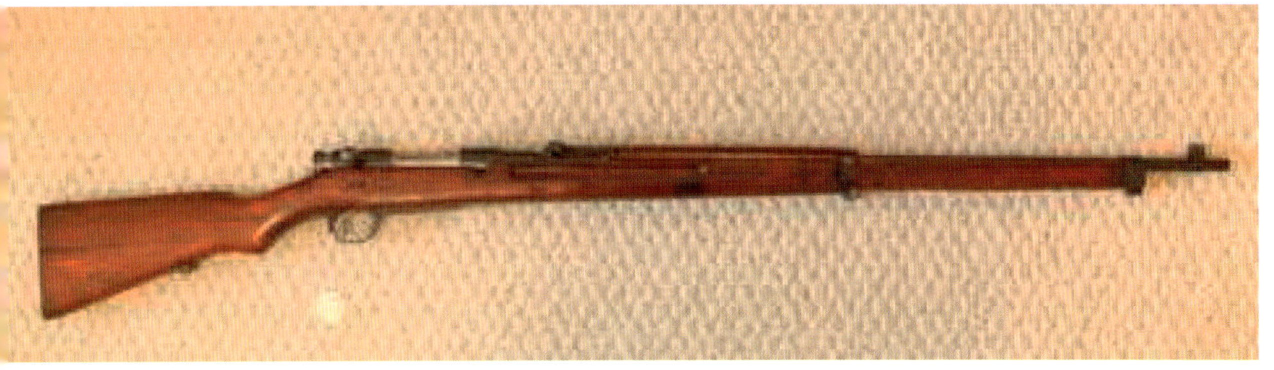

"Muddy" Gentili's rifle was obtained in a trade with some Marines. He was a parachute rigger aboard CV 50 USS Bennington (an aircraft carrier) and has carved all of the places in which the ship was engaged on the stock.

Muddy Gentili in WWII

Abe told me that a surrendering German infantry man ran up to his tank and gave him the gun. The orders were to shoot the Church steeple out because that is where the snipers were. When they shot at the church steeple all the Germans in that town surrendered.
He had also worked as a translator earlier in the war because he spoke Yiddish, which is similar to German.

Abraham Lastnik
U.S.A. 9th Armored Division
2nd Tank Bn. 2nd Battalion, Co. A
1945

This 1905 dated pistol was surrendered to a U.S. tanker by a German infantryman in 1945

This T 30 Arisaka rifle was obtained by Bob Erickson, a motorman on LST 932, when the LST debarked troops from the Lingayen beachhead in the Philippines.
Timing would have been about the 20th of February, 1945. The rifle is a typical T 30, what is not typical is that this then obsolete arm found its way to the PI, was captured, the mum ground, and brought back to the USA by the Veteran. I bought the rifle from the son of the Veteran, Bob Erickson, Jr., same name as is stamped on the rifle, Bob Erickson. Bob, Jr. was kind enough to include a picture of the LST and crew, including his dad; along with a history of the ships service.
The pictures show the rifle and the stamped name. The letters are stamped into the right side of the stock, below the action- fairly typical of Naval/Sailor stuff, as they had access to machine shops and stamps and stuff, not like the GIs. I have no doubt that Bob probably traded something for this souvenir, I doubt he picked it up himself. The mum is ground, which is a little out of the ordinary for a T 30, so it is probably a legitimate pick up.

Explaining why an obsolete rifle was in Lingayen, picked up as a souvenir, then the mum ground is beyond me?
There is also a history of the boat, with the particular page relating to Lingayen shown below.

The LST 932 was at Lingayen twice, and Bob probably obtained the rifle when they backloaded elements of 6th Army Combat Engineer Bn, on 20 Feb. 1945. It is interesting to note that Bob spent his 20th birth day enroute to Lingayen on Feb 11, 1945.

THE HISTORY OF THE U.S.S. L.S.T. 932

1. The LST 932 was built by the Bethlehem Shipbuilding Corporation of Hingham, Mass. She was placed in full commission and under the command of Lt. W. F. JEROME, USNR on 15 August 1944. After the usual shakedown period, she was ordered to load and proceed to the Southwest Pacific and there to report to the SEVENTH Fleet for duty. She reported to the Com7thFlt for duty on 4 October 1944.

2. From date of commission to present date, the ship has been underway 5,291 hours out of a total of 11,640 and has covered a distance of 41,175 miles. She has hauled a total of 21 payloads. Sh was in the D plus 4 echelon at Lingayen and three other D-Day operation in the Philippines. She took part in the Okinawa Campaign, the occupat of Korea and is presently engaged in the repatriation of Japanese and Chinese Nationals. She underwent a torpedo (submarine) attack on 11 February 1945 and has been under attack from suicide planes and bombings in the Lingayen and Okinawa operations. The ship has always been ready for sea on sailing date and on several occasions has substituted for ships who were not ready. The original propellers are still on the shafts and the ship is still operating its original two small boats after 16 months of hard usage.

3. The ship's crew was trained as a unit at Camp Bradford, Virginia and while there was designated as crew No. 4528. This crew, I believe, has been an exceptionally good one, due, in part, to processing at Camp Bradford. In nearly every case, the men were well chosen for the duties they were to perform. Morale has always been high There have been no casualties or even serious accidents. Only three men were lost from the crew for medical reasons. One man was transferred to shore duty because of Chronic Seasickness, another man while undergoing an operation for appendicitis was found to have other complications and was returned to the States for treatment and the third man accidentally shot himself in the hand while standing a security watch. There have been two Summary Courts Martial and 15 Deck Courts Martial held during this period.

4. Attached hereto is a chronological table of the movements of the LST 932 during the period of this history.

U.S.S. L.S.T. 932

CHRONOLOGICAL HISTORY (CONTINUED)

13 DEC. 1944 - ANCHORED HUMBOLDT BAY, NEW GUINEA.

29 DEC. 1944 - UNDERWAY FROM HUMBOLDT BAY, NEW GUINEA FOR SANSAPOR

1 JAN. 1945 - ARRIVED AT SANSAPOR, NEW GUINEA. LOADED FOR MIKE I (LINGAYEN) OPERATION.

3 JAN. 1945 - UNDERWAY FROM NEW GUINEA FOR LINGAYEN, LUZON, PHILIPPINES. LOADED WITH ARMY VEHICLES, PERSONNEL AND EQUIPMENT.

12 JAN. 1945 - ATTACKED BY UNCERTAIN NUMBER OF JAP SUICIDE AIRCRAFT, NO DAMAGE OR CASUALTIES TO THIS SHIP.

13 JAN. 1945 -
* 0700 - SIGHTED JAP PLANES, DRIVEN OFF BY AA FIRE.
* 1200 - ANCHORED IN LINGAYEN GULF. THIS IS D PLUS 4 DAY. HEAVY BOMBARDMENT BY DD'S, CL'S & BB'S.

14 JAN. 1945 - BEACHED ON WHITE BEACH, LINGAYEN, LUZON - DISCHARGED CARGO

15 JAN. 1945 - GOT UNDERWAY FROM LINGAYEN FOR LEYTE.
20 JAN. 1945 - ANCHORED SAN PEDRO BAY, LEYTE.
28 JAN. 1945 - COMLSTGR64 ABOARD. SHIFTED HIS FLAG TO LST 932, THIS SHIP, TO ACT AS TEMPORARY FLAGSHIP. UNDERWAY FROM SAN PEDRO BAY, LEYTE FOR MOROTAI ISLAND.

1 FEB. 1945 - ARRIVED MOROTAI, HALMAHERA GROUP.

8 FEB. 1945 - LOADED ELEMENTS OF 534th EB&S REGIMENT.

9 FEB. 1945 - UNDERWAY FROM PITOE, MOROTAI ISLAND FOR LINGAYEN GULF, LUZON.

11 FEB. 1945 - ENROUTE TO LINGAYEN LST 577 OF THIS CONVOY WAS TORPEDOED AND SUSTAINED CASUALTIES. LST 577 WAS ABANDONED AND EVENTUALLY SUNK BY GUNFIRE FROM ESCORTS.

18 FEB. 1945 - ARRIVED LINGAYEN GULF, LUZON.

20 FEB. 1945 - WHILE BEACHED, JAP PLANE APPROACHED AT 2319, FROM A LOW ALTITUDE. DROPPED 500 POUND BOMB, WHICH EXPLODED 300 YARDS OFF OUR STARBOARD BOW, CAUSING A FIRE. A SMOKE SCREEN WAS LAID. NO DAMAGE OR CASUALTIES TO THIS SHIP. BACK LOADED ELEMENTS OF 6TH ARMY COMBAT ENGINEER BN.

23 FEB. 1945 - UNDERWAY FOR MINDORO.

25 FEB. 1945 - BEACHED ON WEST SIDE OF MINDORO, MANGARIN BAY. DISEMBARKED CARGO & PASSENGERS.

Bob Sr. also "collected" some coins from the Philippines. These supposedly came from a bank safe that was blown open, and the contents distributed; interesting that two of the coins are Chinese, one Japanese.

Not a lot of story, but the paper trail was great, so here is how I picked up this Tokarev.

My buddy and I were setting up at a local gun show. While cruising the hall he picked up a Chinese Tokarev captured in Vietnam and accompanied by trophy papers. The slide is from a T54 while the frame appears to be from an M20. The slide serial # is what's recorded on the docs. She's rough on the outside but the bore is in excellent condition.

Story by Len S told to Merle Fitzgerald

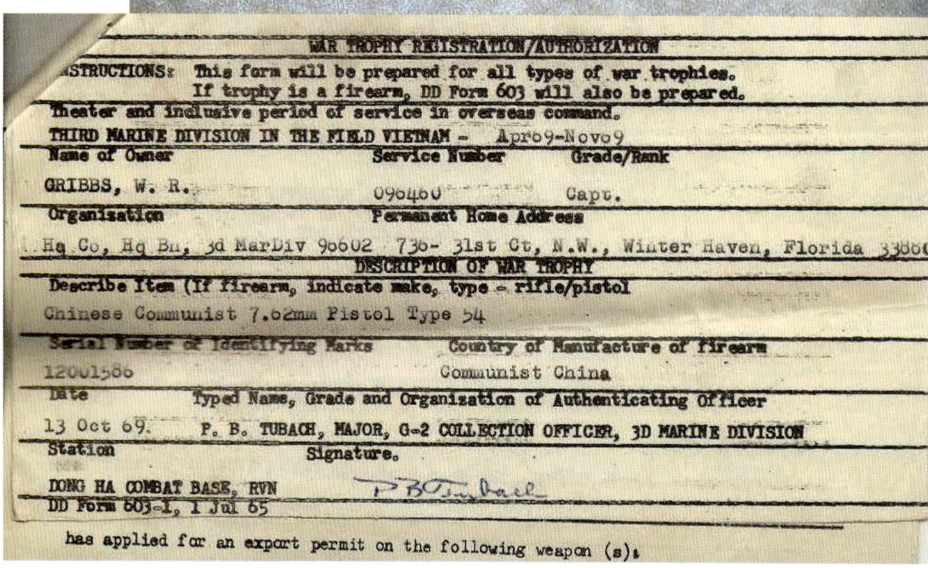

WAR TROPHY REGISTRATION/AUTHORIZATION

INSTRUCTIONS: This form will be prepared for all types of war trophies.
If trophy is a firearm, DD Form 603 will also be prepared.

Theater and inclusive period of service in overseas command.
THIRD MARINE DIVISION IN THE FIELD VIETNAM - Apr69-Nov69

Name of Owner / Service Number / Grade/Rank
GRIBBS, W. R. / 096460 / Capt.

Organization / Permanent Home Address
Hq Co, Hq Bn, 3d MarDiv 96602 / 736- 31st Ct, N.W., Winter Haven, Florida 3388[illegible]

DESCRIPTION OF WAR TROPHY

Describe Item (If firearm, indicate make, type - rifle/pistol
Chinese Communist 7.62mm Pistol Type 54

Serial Number or Identifying Marks / Country of Manufacture of firearm
12001586 / Communist China

Date / Typed Name, Grade and Organization of Authenticating Officer
13 Oct 69. / P. B. TUBACH, MAJOR, G-2 COLLECTION OFFICER, 3D MARINE DIVISION

Station / Signature.
DONG HA COMBAT BASE, RVN / P B Tubach

DD Form 603-1, 1 Jul 65

has applied for an export permit on the following weapon (s):

VIỆT-NAM CỘNG-HÒA

Mẫu số 3

BỘ NỘI-VỤ

Số 10.135 /BNV/KS /16

GIẤY PHÉP GỞI SÚNG VỀ HOA-KỲ

(FIREARMS EXPORT LICENSE)

Nay cho phép W.R. GRIBBS

Dai Uy QDHK

ngụ HqCo. Headquarters Battalion, 3rd Marine Division Saigon.

gởi về Hoa-Kỳ:

— Một cây súng luc hieu Chicom k.k. 7.62 so 12001586

Tịch thu của V.C.

— Phương tiện: mọi phương-tiện

— Ngày gởi phỏng chừng: 20.11.69

Giấy phép nầy có hiệu-lực một tháng kể từ ngày ghi trên.

Saigon, ngày 18 tháng 11 năm 196 9

T.L. Tổng-Trưởng Nội-Vụ

TONG THO KY

TON THAT CHUOC

PROVOST MARSHAL OFFICE

OFFICIAL

USAHAC APO 96243

VIET-NAM

CLEARED FOR CONUS

DEPARTMENT OF THE ARMY

OFFICE OF THE PROVOST MARSHAL

US ARMY HEADQUARTERS AREA COMMAND

REGISTRATION BRANCH

APO 96243

OFFICIAL BUSINESS

POSTAGE AND FEES P

DEPARTMENT OF THE A

US AIR MAIL

CPT W.R. GRIBBS

736 31st Ct. Northwest

Winter Haven, Florida

33880.

REGISTRATION OF WAR TROPHY FIREARMS

The original of this registration form will be retained by the person authorized possession. This registration is not transferable.

1. THEATER: HqCo., Headquarters Battalion, 3rd Marine Division(REPUBLIC VIET NAM)

2. NAME OF OWNER (Last name, first name, middle initial): GRIBBS, W.R.

3. SERVICE NO.: 096460

4. GRADE: Capt.

5. ORGANIZATION: HqCo., Headquarters Battalion, 3rd Marine Division(REIN)FMF FPO SAN FRANCISCO 96602

6. PERMANENT HOME ADDRESS (Street, city, State): 736 31st Ct., N.W., Winter Haven, Florida, 33880

7. DESCRIPTION OF FIREARM

a. MAKE	b. TYPE (Rifle, pistol, shotgun, etc.)	c. MODEL
Chicom	Pistol	type 54
d. SERIAL NO.	**e. CALIBER**	**f. COUNTRY OF MANUFACTURE**
12001586	7.62mm	CHINA

8. DATE: 22 Oct 69

9. TYPED NAME, GRADE, AND ORGANIZATION OF EXECUTING OFFICER: C.W. MURRAY JR, 1st Lt, OPERATIONS OFFICER, 3rd MarDiv PMO

10. STATION: QUANG TRI COMBAT BASE

11. SIGNATURE OF EXECUTING OFFICER: C.W. Murray Jr

DD FORM 1 OCT 51 603

REQUEST FOR FIREARMS EXPORT LICENSE

22 Oct 69 (date)

Name (First, Middle, last) W.R. GRIBBS,

Occupation (Title, rank) Capt. Ser. No. 096460

Agency USMC Agency/Unit address HqCo., Headquarters Battalion, 3rd Marine Division

Date of Departure 5 Nov 69

Airline n/a Flight No. n/a

Type of firearm (pistol, revolver, rifle, shotgun) Pistol

Make of Firearm Chicom

Caliber (m/m) 7.62mm

Serial number of firearm 12001586

Vietnamese gun permit No. n/a Date of issuance n/a

The weapon described above will be hand carried or shipped via accompanied Baggage on or about 5 Nov 69 (date).

In the event I make arrangements to dispose of the weapon described prior to my departure from Vietnam I will immediatly notify the Provost Marshall, USAHAC, Saigon so that the requested export license may be cancelled.

W.R. Gribbs

Signature of Applicant

USARV Reg No. 643-20

TEMPORARY EXPORT LICENSE

OFFICE OF THE PROVOST MARSHAL
THIRD MARINE DIVISION (REIN) FMF
FPO SAN FRANCISCO, 96602

3G-27

22 Oct 69

SUBJECT: Transporting Firearms

TO: Chief, Vietnamese Customs

1. This is to certify that Captain W.R. GRIBBS, 096460, HqCo., Headquarters Battalion, 3rd Marine Division has applied for an export permit on the following weapon (s):

   Type of firearm Pistol

   Make of firearm Chicom

   Caliber (MM) 7.62mm

   Serial number of firearm 12001586

2. Request permanent export license be mailed to 736 31st Ct., N.W., Winter Haven, Florida, 33880

CLEARED FOR CONUS

C.W. MURRAY JR
By Direction

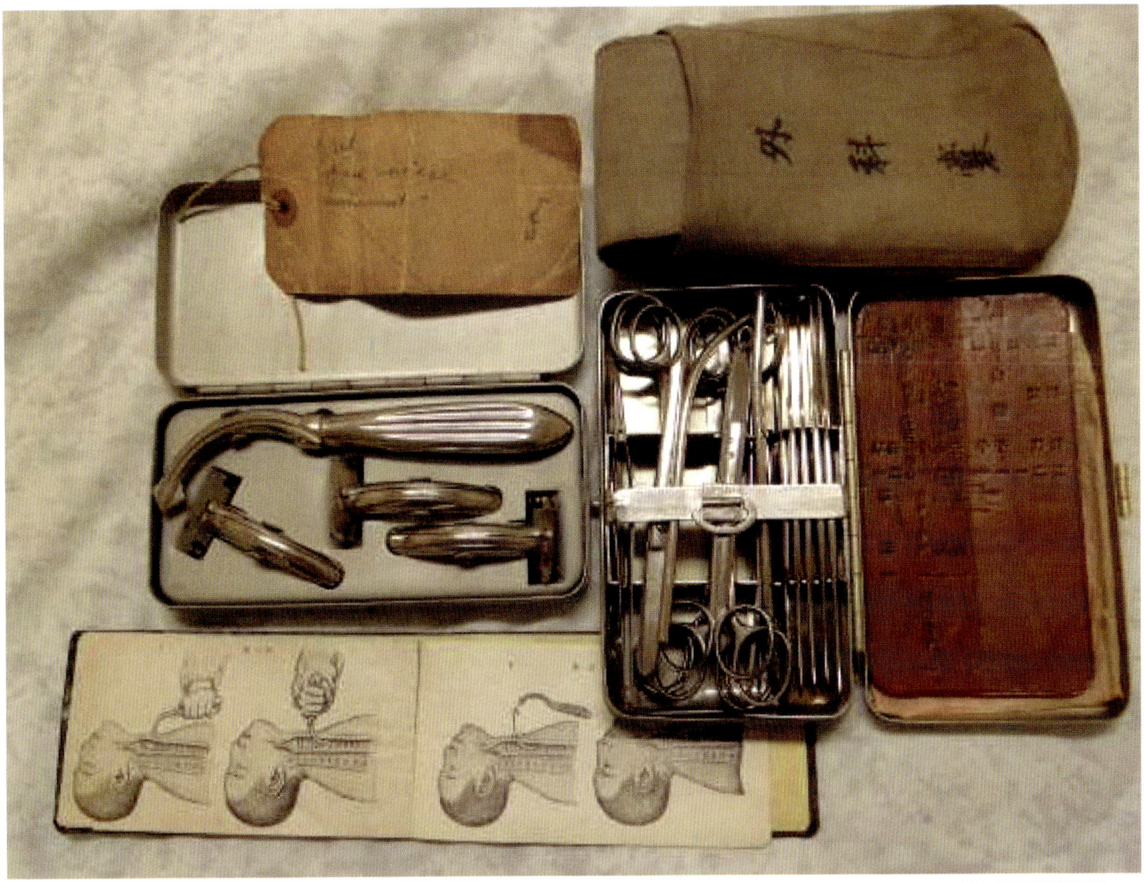

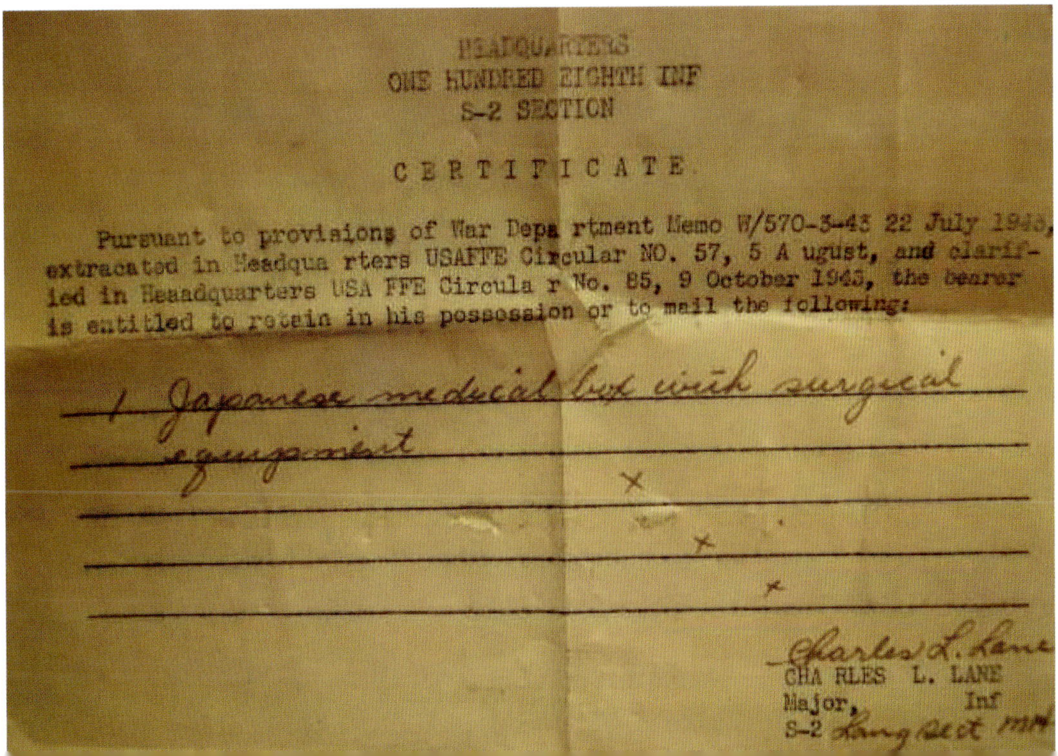

HEADQUARTERS
ONE HUNDRED EIGHTH INF
S-2 SECTION

C E R T I F I C A T E

Pursuant to provisions of War Depa rtment Memo W/570-3-43 22 July 1943, extracated in Headqua rters USAFFE Circular NO. 57, 5 A ugust, and clarif-ied in Headquarters USA FFE Circula r No. 85, 9 October 1943, the bearer is entitled to retain in his possession or to mail the following:

1 Japanese medical box with surgical equipment

Charles L. Lane
CHA RLES L. LANE
Major, Inf
S-2

This is a photo of my Dad, Robert L Rhodes on Okinawa with a Arisaka T-99 rifle he picked up. I'd give anything in my collection to have this rifle. Dad gave it to his cousin who lost it!
Dad was in the Army Engineers and served on Saipan and Okinawa.

First published in *The Japanese Type 99 Arisaka Rifle* 2010 edition by Don Voigt

About 35 years ago, a widow was having a sale of some of her deceased husband's things. Due to the time period I do not remember his name. One of the items was a 1934 Mauser - Navy marked, along with a navy marked holster and a sack. Her husband was in Bremen after the end of the war but before the British pulled out. Like all service men, the Brits and the Americans drank and gambled. It was during one game of chance that had been going on for four days (according to the widow), that a Brit put the pistol and accessories into the pot. He lost. The American ended up with the rig.

The Brit had been one of the first into the weapons room at the naval facility and picked it up from the rack.

The veteran was soon sent back home and he still had the pistol rig. Afterwards, he loaded the pistol, and put it and the holster under the cushion on his chair in the living room - in case he might need it. It was still there when he died. The widow was moving in with one of her daughters and had no further need for it. The pistol had never been fired by the Brit or the deceased and is in 100% new condition !

Navy waffen marked holster

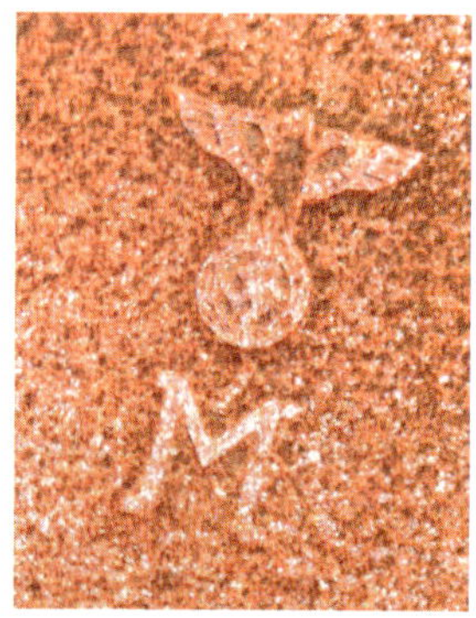

US Army, 9th Infantry Division & MACV
1970 - 1971
Mekong Delta
Warrant Officer One, James Jackson

The Tokerev Pistol & Holster were taken off an enemy sapper that attempted to breech their perimeter one night. He was armed with an RPG, satchel charges and this pistol. Needless to say, he did not make it. Note the back of the holster; it was altered by the Vietcong, by adding additional loops to hook on a web belt. The pistol is Chinese and completely matching.

My dad gave these to me a year ago as a gift and made me promise to leave them to my children.

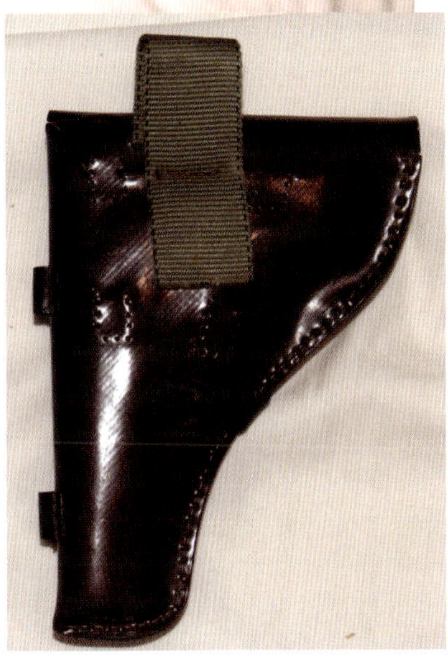

This WWII era, Italian Carcano my dad picked it up at a weapons cache they found in a local village. He stated he had never seen a rifle like it before and most of the rifles were M44s and he grabbed this one as it was one of two in the entire cache. They were not allowed to bring full-autos home, so everyone grabbed a bolt action rifle or an SKS.

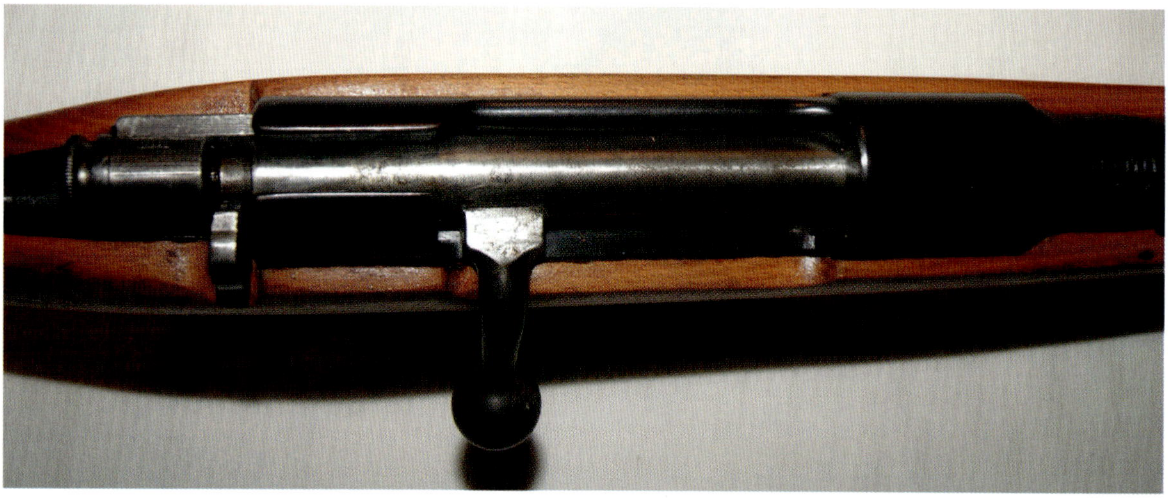

My Uncle Tennie Clarin (1920-1998) served in the 90th Infantry Division, 344th Field Artillery Battalion during WWII. This division was a part of General Patton's Third Army.

A TRIBUTE TO THE 90TH DIVISION:
"Sometimes I think you don't know how good you are. You are the best soldiers in the world. It was a great honor to command you." General George S. Patton, Weiden, Germany, July 13, 1945

- The 90th had more days of combat than any other Division in the ETO.
- The 90th had at least one unit in contact with the enemy every day from D-Day to VE- Day.
- The 90th earned the Invasion Arrowhead and all five Battle Stars for Normandy, Northern France Ardennes [Battle of the Bulge], Rhineland Central Europe, and also the French Croix de Guerre.

• The 90th crossed dozens of rivers, liberated hundreds of towns and several million people. The 90th liberated Flossenberg Concentration Camp and captured the Merkers Salt Mine intact with all of Germanys store of gold and art treasures
• The 90th forced the surrender of the German 11th Panzer Division en masse and met the Russians in Czechoslovakia
• The 90th suffered over 15,000 battle wounds. The total number of men treated by medics to include all injuries (trench foot and frostbite) was 25,988.

Tennie Clarin was a combat medic and also drove a truck that carried the medical supplies when the unit was on the move. The truck had an open cab with only a flip-up windshield for protection. My Uncle along with the other Americans had to endure terrible conditions and one of the worst winters ever as they battled their way across France, Germany, and even into Czechoslovakia. The 90th Division was also at Zella-Mehlis, Germany in early April, 1945 where they helped capture and occupy the town. This town was home to the famous Walther arms factory. He acquired a P.38 pistol, ac 45 serial number 181c. He was an avid outdoorsman his entire life and enjoyed fishing and hunting.

Contributed by Ron Clarin

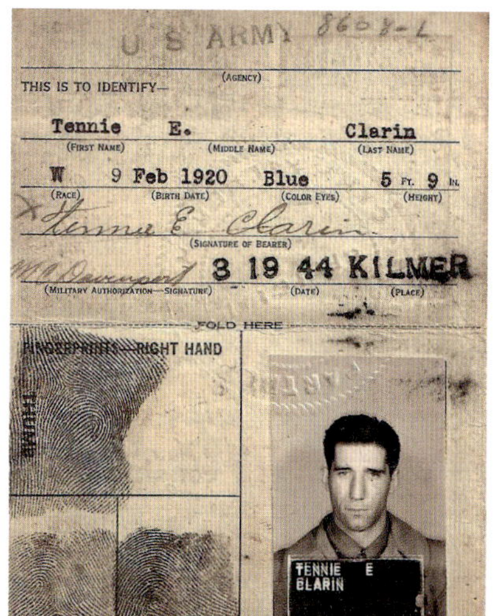
U S ARMY 8608-L
(AGENCY)
THIS IS TO IDENTIFY—
Tennie E. Clarin
(FIRST NAME) (MIDDLE NAME) (LAST NAME)
W 9 Feb 1920 Blue 5 FT. 9 IN.
(RACE) (BIRTH DATE) (COLOR EYES) (HEIGHT)
(SIGNATURE OF BEARER)
3 19 44 KILMER
(MILITARY AUTHORIZATION—SIGNATURE) (DATE) (PLACE)
FOLD HERE
FINGERPRINTS—RIGHT HAND
INDEX MIDDLE
TENNIE E
CLARIN
10—28605-1

This is a 91/30 sniper rifle brought back by Marine Sgt Lonnie Patten, who also brought back a Makarov. Lonnie was a collector, and brought back many weapons (I also have a Tokarev pistol from Lonnie) as well as uniforms, belts, medals, etc. This sniper rifle came with the following history as related:

"The sniper rifle was taken while I was on detached duty from the 5th Marines, serving in a CAP (Combined Action Platoon) in Quan Nam Province at Duc-Duc and Duc Thaim district in 1969. The NVA sniper team had been creating havoc with the troops of the 5th and 11th Marines around the An Hoa Combat Base. Several major attempts had been mounted against them by conventional search methods from the 5th Marines, but the sniper team eluded all operations against them.

Our CAP unit received an intelligence report that the team would be in our area to terminate the USMC presence and show the NVA power in the area. We set up an ambush site in an area we figures would be a good sniper spot. After a day wait, the sniper team consisting of two individuals with three support troops. One VC and two NVA came into the ambush site at which time the ambush was sprung on them, killing all five. Photographs were made, documents and identification were taken. The sniper was of Eurasian stature while his teammate was Chinese and the three support troops were Vietnamese."

I live in Spain and my P08 has been fired many times at practice sessions, just not recently. It was first used by my wife's grandfather who got it as a present from a German officer on an inspection tour to the Spanish "Blue Division" on the Eastern Front. At the time he was a staff Major in the Spanish army, and then it was passed on to my father in law (retired army Brigadier) when he graduated as a lieutenant. He used it all his active life until he retired 10 years ago. Spanish officers had the choice of either being issued a standard pistol or buying their own.
Its 100% original finish, with all parts matching. Also I was given two original magazines with wooden stops, sadly it didn't come with the original holster. All maintenance it has undergone has been done at their respective unit's armory with great care as it was the "general's personal pistol."

1917
ERFURT
21

4621
21
21

4621

Pacific Theatre approximately 1944 or 1945

This Chinese military Makarov was brought back by Lonnie Patten, who was in the 5th Marines. He captured the pistol in 1970 from a North Vietnamese Army (NVA) Colonel during an ambush operation in I Corps near the hamlets of Duc Thaim in the Arizona Territory. Lonnie stated that there had been an increase in kidnappings, assassinations, and executions in the area by due to the arrival of a new NVA Colonel. Intelligence also noted that he had relatives in that area, and the NVA travelled freely, and that he had given orders for the kidnapping of several German doctors and nurses. With ARVN and Combined Action Company personnel it was arranged to ambush the NVA colonel when he crossed the Song Vu Gia River on one of his village visits. The ambush was successful and this pistol was captured.

My grandfather joined the army, right before WWII began; made Sergeant and then attended OCS and came out a 2nd Lieutenant. After Pearl Harbor was attacked, he was assigned to supervise the radar sites. In a talk with him several years ago, before he died, he told me that it was actually a very trying time. No one was sure if the Japanese would attack again, since the fleet had not been destroyed as hoped. He served through WWII on Oahu, Hawaii. He did pick up an Arisaka, but it was from a friend of his in the 1950's. Contributed by Edward Tinker

Photos submitted by Sandra Herbig Tinker

Taken at the position on May 15, 1942. Not posed for either. More or less caught with the "Pants down". Notice the missing belt and thoughtf attitude.

During WWII, my Grandmother moved to Hawaii to be with him. She told my mom, that he drove her around to many hill sites to show her what he did. I assume it was in a vehicle like this. In 1984 my grandmother visited and we found their old house; most older structures have been torn down. You can still see Zero machinegun pock-marks on the barracks at Schofield Barracks, Hawaii.

My father David W. Caswell was a Fire Control man 2nd class onboard LST 1062. In the fall of 1945 his ship brought back Japanese prisoners to Japan, stopped in Okinawa and conducted occupation duty in Japan..

Isn't this color photo of my father just great!

On the LST 1062

Japanese Battle-ship Nagato

Family picture before heading out

My fathers ship brought back Japanese prisoners. Here are assorted pictures of the prisoners, some literally washing up, others being fed on deck.

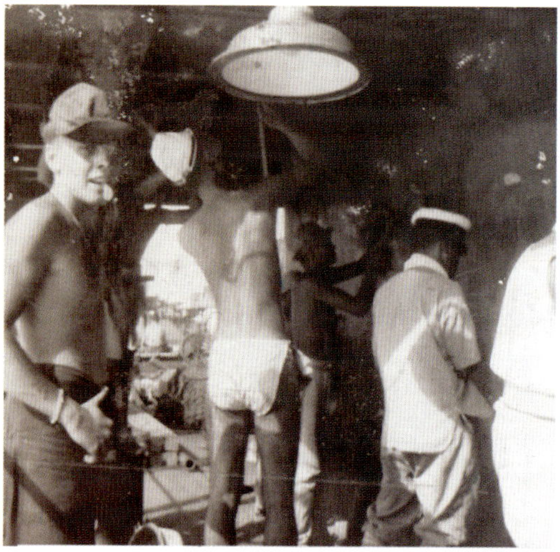

Occupation duty in Japan

Emperors Palace

These are from occupation duty. His ship bent a screw in a typhoon in September 1945 and had to go to Yokasuka (?) to be dry docked for repair.

Some are a bit touching..

Lagoon next to the Emperors Palace

2nd Class Fire Control Man David W. Caswell with assorted pictures of damaged aircraft, US vehicles and some ships from Okinawa to occupation duty in Japan.

Japanese Black Widow

USS New Jersey

USS Missouri

30 Caliber "IRWIN PEDERSEN" Carbine Rifle

It was in 1943, when Lt. Robert Lynch issued this rifle, serial # 1768802, to me. We were in the South Pacific at the time. What makes this particular rifle so valuable and special is, that according to Internet research, only a total amount of approximately 4000 "IRWIN PEDERSEN" M 1, 30 caliber carbine rifles were manufactured. As far as I know, this rifle is in its original condition. Nothing has been altered and no parts have been exchanged, except for the addition of the shoulder strap and magazine pouch and the butt oiler.

I was a Radio Technician 1st Class on board the U.S.S. El Paso, a patrol frigate P.F. 41 in the United States Coast Guard. We were landing about 12 Army Rangers on the coast of Peleliu Island to mop up. As the Rangers were approaching the Island, the Japanese Army fired small arms at us, and we were returning fire with 3" guns and 40mm bo-fors canons and other small arms to cover our Rangers.

At the end of the war, when the U.S.S. El Paso was in the Brooklyn Navy Yard for repairs, I told Lt. Lynch, that I would like to take my rifle home. All enlisted men - except for officers - were searched, leaving the Naval Base. Lt. Lynch offered to take the rifle ashore for me and then handed it to me, and I then mailed this rifle home to California in a canvas bag.

Robert Davis
....

Robert Davis, born March 10, 1917

David Barnette—Rare M1 Carbine

Mr. Flagg was a cook on board a transport vessel that had stopped in Saipan. He and some crew mates went ashore and came across a cache of captured weapons. From what we gathered the cache was unguarded and they all proceeded to take a rifle. He took the rifle back to the ship and eventually sent it home. There aren't any official papers for the rifle. Somewhere along the line someone must have taken out the firing pin and stuck the bolt in backwards. As you can see from the pictures it is still missing some of the bolt parts and the bolt is lodged in there backwards. The rifle had remained in his closet since it came home and his family wasn't even aware of it until about 2 years ago. When they found it they wanted to know more about it and this year they finally decided to go to a gun show to get some more information. That is when I met them and talked to them about the gun. It came out that they were on hard times and needed to sell it. I offered to buy it from them as long as I could document their story. This is the basic version of the story but there will be more information soon. Now I have to clean the rifle up. it has lots of surface rust and has some splotches of pink paint on it.

Mr. Flagg had taped some papers with his name on it so I don't know how I can take it apart without ruining the papers.

A couple of US GIs playing around circa 1920

Dad's name was Alfred Nielsen aka Corporal Alfred Nielsen and he served throughout the European theatre. He did what a lot of folks did, picked up some souvenirs for himself. However, Alfred got and kept a bit more than many folks and his daughter Diane has shared them with us! (There is a previous story of his Luger capture in the book). Here is a copy of a letter he sent home:

Cpl. A. Nielsen 31326465 (246) Postmark: June 9, 1945
Hq. Btry. 740th F.A. Bn.
A.P.O. #403
C/O Postmaster New York, New York

June 8, 1945
"Germany"

Dear Mom & Pop: -

Well, here's your Busse again with some line to "youse". First off, how's every little thing at #162? O.K. I hope 'cause all' on the up & up over here with me. The weather has been swell for the last couple of days without rain at all.

I sent a box home yesterday with a lot of odds & ends that I have picked up over here. Here's a list of the stuff that's supposed to be there:-

2 - Nazi party daggers
1 - bayonet
1 - Iron Cross Medal
12 - pins
2 - small Nazi flags
1 - German belt buckle
5 - Foreign coins
1 - piece of V-bomb
2 - Nazi arm-bands
1 - Nazi insignia
1 - propaganda leaflet
1 - Stars & Stripes (souvenir)

All this stuff is just a lot of junk but who cares --- almost everyone has a story behind it & I'll tell you all about it when I get home. You will probably notice that the box was sent to me by a "Miss Gene Miller" --- really that's Marion Thompson. You see, she sent me a quart of "Four Roses" & in case it broke she wouldn't be blamed. Tricky, ain't she?

I got your letter last nite with Farmor's address in it and that sure was a swell idea about me writing up there --- I had forgotten all about that Jytte can read English, so tonite I'll sit right down and write them a long letter about us with new house & all just in case they don't know about it.

Well, this is all for now so I'll just say--- So Long & be good.

Your loving son,
Al

The former Lieutenant, now a retired major, no longer wanted to have the gun that had shot him, and I agreed to purchase it from him. With the rifle came the statement, a copy of his Purple Heart orders, many photos, and a friendship. One very interesting side note was that I asked him if he had known a high school friend of mine who had been KIA while serving as platoon leader of the Recon Platoon with 5/60. It turned out that as lieutenants they had been good friends, and he had listened on the radio to the action where my friend was killed. He told me of the profound sadness that permeated after my friend and several other Recon Platoon members died that day.

I treasure this rifle as a link to my new friend as well as to my old high school friend.

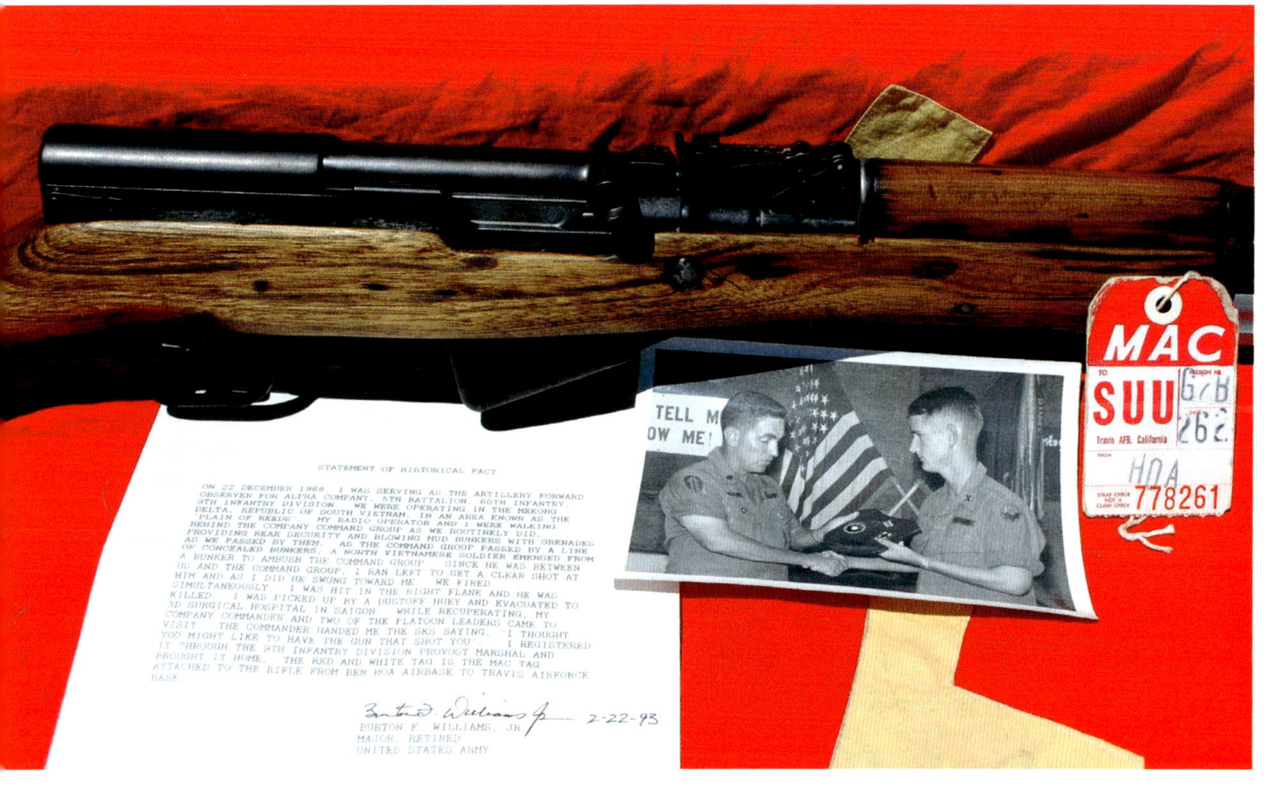

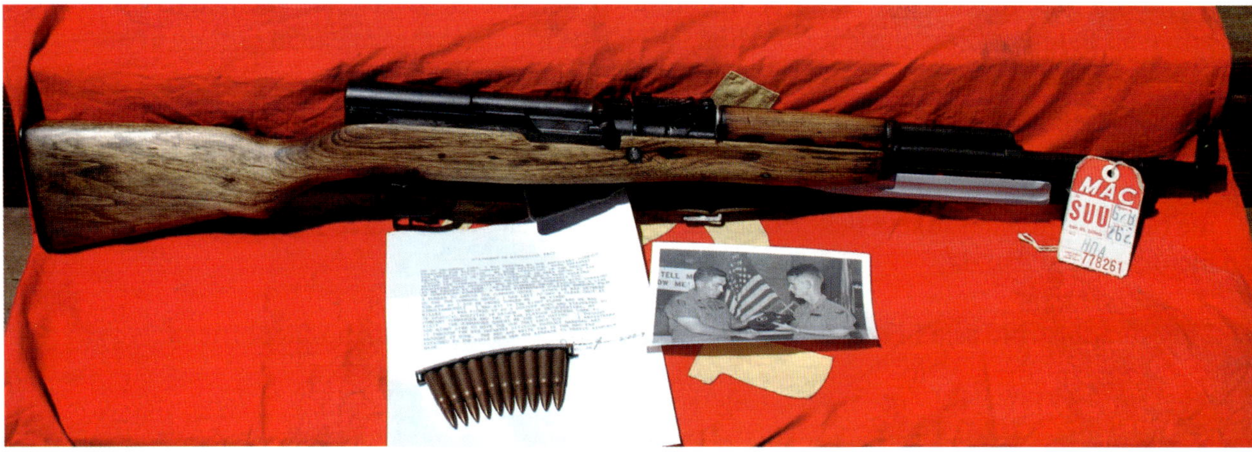

STATEMENT OF HISTORICAL FACT

ON 22 DECEMBER 1968, I WAS SERVING AS THE ARTILLERY FORWARD OBSERVER FOR ALPHA COMPANY, 5TH BATTALION, 60TH INFANTRY, 9TH INFANTRY DIVISION. WE WERE OPERATING IN THE MEKONG DELTA, REPUBLIC OF SOUTH VIETNAM, IN AN AREA KNOWN AS THE "PLAIN OF REEDS". MY RADIO OPERATOR AND I WERE WALKING BEHIND THE COMPANY COMMAND GROUP AS WE ROUTINELY DID, PROVIDING REAR SECURITY AND BLOWING MUD BUNKERS WITH GRENADES AS WE PASSED BY THEM. AS THE COMMAND GROUP PASSED BY A LINE OF CONCEALED BUNKERS, A NORTH VIETNAMESE SOLDIER EMERGED FROM A BUNKER TO AMBUSH THE COMMAND GROUP. SINCE HE WAS BETWEEN US AND THE COMMAND GROUP, I RAN LEFT TO GET A CLEAR SHOT AT HIM AND AS I DID HE SWUNG TOWARD ME. WE FIRED SIMULTANEOUSLY. I WAS HIT IN THE RIGHT FLANK AND HE WAS KILLED. I WAS PICKED UP BY A DUSTOFF HUEY AND EVACUATED TO 3D SURGICAL HOSPITAL IN SAIGON. WHILE RECUPERATING, MY COMPANY COMMANDER AND TWO OF THE PLATOON LEADERS CAME TO VISIT. THE COMMANDER HANDED ME THE SKS SAYING, "I THOUGHT YOU MIGHT LIKE TO HAVE THE GUN THAT SHOT YOU". I REGISTERED IT THROUGH THE 9TH INFANTRY DIVISION PROVOST MARSHAL AND BROUGHT IT HOME. THE RED AND WHITE TAG IS THE MAC TAG ATTACHED TO THE RIFLE FROM BEN HOA AIRBASE TO TRAVIS AIRFORCE BASE.

Burton F. Williams Jr

BURTON F. WILLIAMS, JR.
MAJOR, RETIRED
UNITED STATES ARMY

2-22-1993

Jap 7.7 Arisaka rifle picked up on Guam July 29, 1944 by PFC Harold Kelly, 3rd Marine Div.
Damage is from grenades thrown into cave

James Weldon Lewis

Despite being deferred from military service because his work as a heavy-equipment operator was considered essential for the war effort, the 22-year-old Iowa native enlisted in the U.S. Navy on August 8, 1942, at Norfolk, Va. After Navy basic training, Weldon volunteered to serve as a combat corpsman (medic) attached to the U.S. Marine Corps and was transferred to the Marines. (Note: The Marine Corps is technically a branch of the Navy and uses Navy medical personnel.) In any case, lucky Weldon got to once again go through basic training — this time with the Marines.

After training at Parris Island, S.C., and Camp Elliot, Cal., he was assigned to Company D, 2nd Medical Battalion, 2nd Marine Division. Weldon was part of a reserve combat unit on a troop-transport ship off the shore of Betio Island during the battle of Tarawa in November 1943. Because of the overwhelming number of casualties, the troop transport soon became a makeshift hospital ship. Weldon and other Company D corpsmen helped take care of casualties coming directly from the battlefield. More than 1,000 Marines died at Tarawa and nearly 2,300 were wounded in four days of fighting. Only 17 of the more than 4,500 Japanese defenders survived the battle.

Next came the battle of Saipan in the summer of 1944. Weldon's company was attached to the 2nd Marine Division's Combat Team 6, the core of which was the 6th Regiment. He was with the first waves of combat troops to go into "Red Beach" on June 15, the first day of the battle. It was rough — the commander of the 6th Regiment lost nearly all his officers and staff before noon, and more than 1,400 of the nearly 4,000 Marines in Combat Team 6 were killed or wounded before nightfall. That first tough day would be followed by many more.

Weldon was a front-line corpsman throughout most of the battle, which stretched into mid-July. He saw much action. Saipan was a fierce struggle — when it was over U.S. forces had suffered more than 16,500 casualties, of which nearly 3,500 were killed in action. An estimated 30,000 Japanese troops died on Saipan, and more than 22,000 Japanese civilians committed suicide or were killed by Japanese troops to prevent their surrender. The naval and air battle that was part of the Saipan campaign resulted in a crushing defeat for the Japanese, who lost three aircraft carriers and around 600 aircraft.

In late July and early August of 1944 Weldon saw more combat during the battle of Tinian. U.S. forces suffered nearly 2,000 casualties during that fight, and more than 11,000 Japanese troops and civilians died. The island quickly became the hub of the U.S. bombing campaign against Japan. The B29s that dropped the atomic bombs on Hiroshima and Nagasaki flew from Tinian. In 1945 Weldon was transferred back to the Navy and served as the pharmacist's mate aboard the USS YMS 119, a minesweeper patrolling the waters off Pearl Harbor.

As the only medical crewman on the entire vessel and as the ship's only combat veteran, he was well-respected by his shipmates. This veneration made it possible for him to do things such as talk a drunken crewman into giving him a knife just after the sailor had threatened the ship's commander with it. Weldon then talked the skipper out of a court-martial proceeding against the sailor, who was basically a lonely kid who drank too much during a wild night in Honolulu.

He was a member of the U.S. Navy and Marine Corps and left service as a Corpsman with the rank of First Class Petty Officer by the war's end. He earned five medals:
1) Victory — World War II
2) Good Conduct
3) *American Campaign* (as a sailor on a minesweeper ship on patrol out of Pearl Harbor, Hawaii*)*
4) *2nd Marine Division Presidential Unit Citation*
5) Asiatic-Pacific Campaign with three battle stars (as a troop-ship corpsman during the battle of Tarawa, and as a combat corpsman during the Saipan and Tinian battles)

Although he could have stayed in the service and was scheduled for a promotion to Chief Petty Officer, Weldon decided to return to civilian life in late 1945. After a long train ride from the West Coast back to his point of enlistment in Norfolk, Va., he was discharged on December 28, 1945.

During his time with the Navy and Marines, Weldon traveled across the United States more than once and was bivouacked in places from New Zealand to the "Big Island" of Hawaii. He stood review to (and managed to shake hands with) such luminaries as Admiral Chester Nimitz, commander-in-chief of the U.S. Pacific fleet, and First Lady Eleanor Roosevelt. Of most importance, as a Marine combat corpsman he helped save the lives of many young men.

Chuck Lewis—Type 99 short rifle

A Nagoya Series 1 Arisaka Type 99 short rifle, with two Japanese grenade indents on the right butt flat, a Kokura bayonet, a "Kiseru" (Japanese tobacco pipe and teak holder, a pair of "Jikatabi" Japanese lounging/night-fighting shoes, a pair of Japanese medic sulpha containers, one of dad's USMC garrison caps, his new testament, which he carried throughout combat and his USMC lower pack, with D Company 2nd Medical Battalion markings

All of the Japanese items are things he sent home from or brought from Saipan. He said that the rifle he turned in "to Division" was not the one they gave him back to ship home, although it is interesting with those grenade-fuse indentations. The one he turned in had a nice, clean machine-gun bullet hole through the stock. He said "some officer probably grabbed it."

Dad saw lots of combat plus had other interesting experiences. He had guys die in his arms; watched in horror as Japanese civilians committed suicide by jumping off the Marpi point cliffs; lost several close friends; treated more than one guy with a fractured knee because the guy thought a Japanese knee mortar should be fired from the knee; etc. He said the nights were the worst because the Japanese usually attacked. They knew an attack was coming when they would hear those grenade fuses knocking against something hard.

Here's the photo taken of him in combat on Saipan in 1944. It was taken by a USMC combat photographer and later sent to him. The photographer took down his name and address and apparently survived the war. It has been colorized.
He said the pants he has on were way too big for him. He got them somehow while in combat because his original ones were ripped up and blood-soaked....

My Grandfather's name is Elon E Rhodes and he was a Sgt. in the MTC (Motor Transport Corps). I don't have any details on the watch other than he brought it back from France.

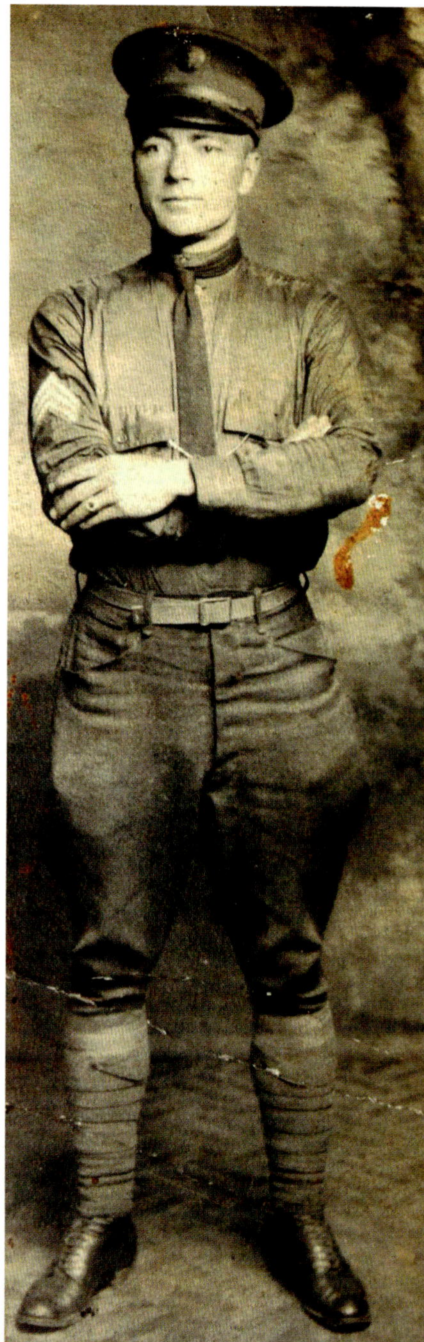

While in Iraq in 2004 I ran across a few of these Tariq pistols in raids and such.
For the most part they were just common plain pistols, with a "bust" of an ancient warrior on the grips. One day I picked up one which had writing on the slide in Arabic, but instead of the "bust" it had the Iraqi crest in the grips. I gave it away (like most of the stuff I ran across) and thought nothing of it. About a month later the guy I gave it to, comes up to me and asks "do you know what was inscribed on that Tariq you gave me." Of course I was all ears, he said it translated into "Presented to a Warrior from the President of Iraq Saddam Hussein."

.

As things worked out, I came across this one. Although it has just the "bust" grips, it also has the inscription. I was told these would be given out to officers as a tribute to a job well done.

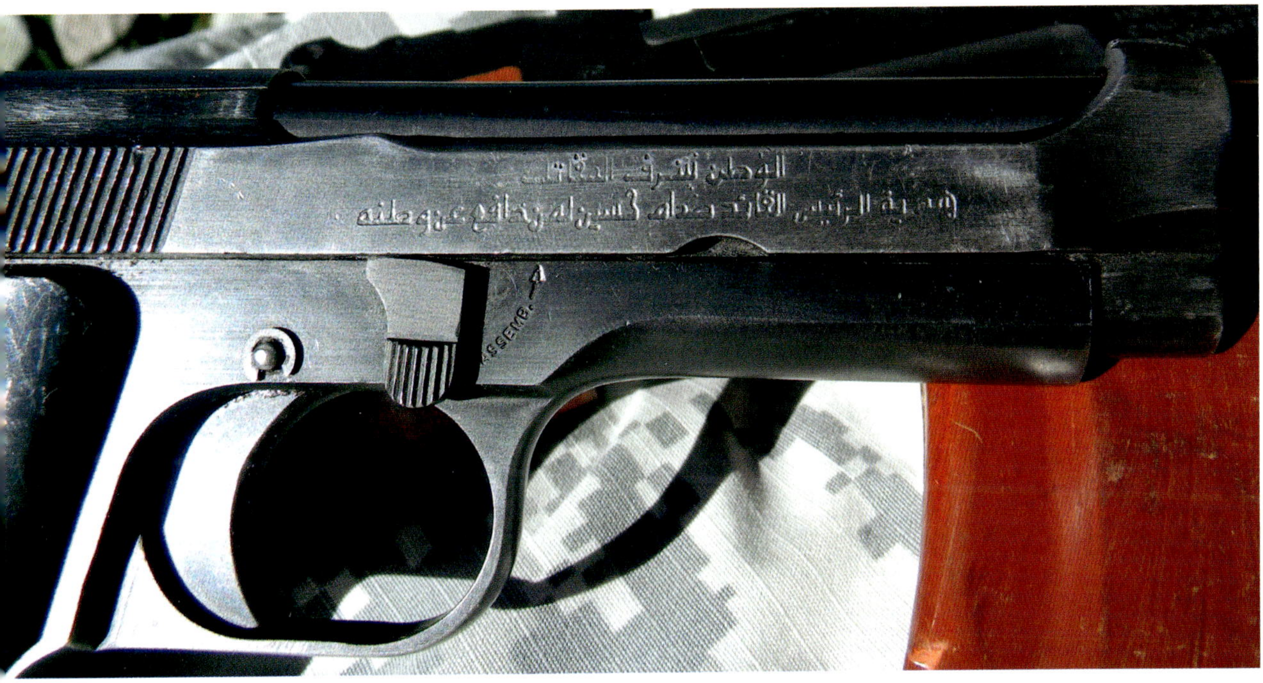

The slide translated says: Defending the country is the soldiers honor

A present from Saddam Hussein to whomever defended his country (translated by a Saudi Friend)

Al—Iraqi AK-47

AK-47—I picked up a lot of different AK's during my time in Iraq, and even used them on occasion. This one I had for about 4 months, it is all matching to include the front hand guard. As I was getting ready to leave, I thought about what to do with it, (we had been through a lot together), and to give it away like all the rest I had or bring it home. Now I was not about to bring back a full auto rifle, who needs that headache. But then I had the idea about bringing back just the parts which were still allowed to be imported into the US. So I cut the receiver off and brought back the parts. I had a lower built, with the required US compliant parts and asked that the serial number be that of the original rifle.

When I got it back I was floored. The gun looks just like it did when I carried it, not only did it have its original Bulgarian serial number it also sported a 10 inside a circle like the original. It was like seeing an old friend.

When I am gone these two will go to a local museum, this way they will continue to serve together.

The Story of Mauser P-38 Pistol Serial # 2567 f

I was a pilot in the 394th Bomb Group, 586th Bomb Squadron flying Martin B-26 Marauders. I joined the group at Cambrai, France in October 1944 and flew 33 combat missions before the war ended. Of these missions our crew was credited with 10- group lead, 7-box lead and 6-flight lead missions.

The group had moved to Venlo, Holland in order to be closer to the front lines; however, the war ended and we never flew any missions from Venlo.

Prior to going overseas, I had been a flight instructor in the Martin B-26, and at the wars end had logged 1,375 hours in the B-26 and held the rank of captain.

At Venlo, we continued to fly training missions. One day, a captain from the infantry arrived on base to visit with his brother who was our squadron first sergeant. He wanted a ride in the B-26, and I was selected to be the pilot.

We took off from Venlo and flew all the way to Berlin following the Autobahn and flying about 200 feet above the ground at about 210 MPH. The captain got a birds eye view of some of Germany and thoroughly enjoyed the flight.

After we landed, he took me to a weapons carrier that had a footlocker in the back, and it was loaded with German pistols, rifles, daggers and medals of all sorts He told me to pick out a pistol, and I chose this P-38 Mauser, serial number 2567 f. He also gave me a German dagger and some medals. He told me that he had gotten the P-38 pistols from the German factory and that they were completely assembled. I doubt that what he was doing was completely legitimate, but I was surely at the right place at the right time.

I have always cherished the pistol. I have fired a few rounds through it at the firing range, and it always operated flawlessly.

I was released from active duty and in Nov. 1946 joined the 196th Fighter Squadron, California Air National Guard when the unit was activated in Nov. 1946. I realized my dream of being a fighter pilot and flew the P-51 D and H, P.47, F-80, F-84E, F-86A, L and H, F-94 A, B and C, F-102 aircraft and the C-47 and the Douglas A-26. The squadron was recalled to active for 21 months during the Korean emergency, and I was stationed in Japan and Korea but flew no combat missions. I retired from the military in Feb. 1969. At that time, I was commander of the l63rd Fighter Interceptor Group, California ANG based at Ontario International Airport, California and flying F-102 aircraft.

I retired with the rank of Colonel and I am a Command Pilot with 7, 845 total hours mostly in single engine jet fighter aircraft.

Donald L. Frisbie
Colonel, U.S.A.F (Ret)

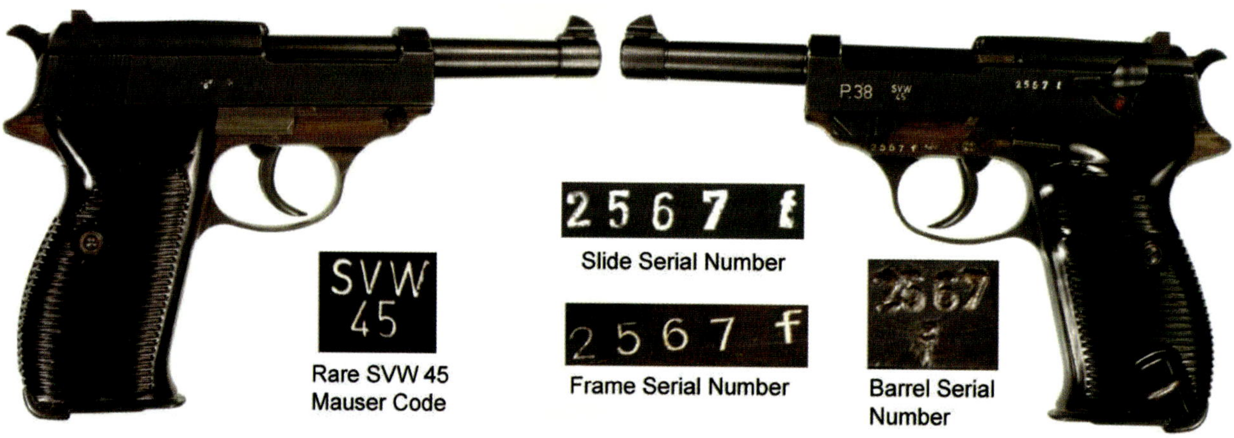

This is a super rare all blue svw 45 not a re-blue but a factory late war all blue. I asked the Veteran why he picked the all blue gun out and he said the other guns in the crate looked like they would rust because the had a bad finish. (Because they were phosphated and he was used to seeing blued guns!)

WAR TROPHY REGISTRATION/AUTHORIZATION

INSTRUCTIONS: This form will be prepared for all types of war trophies. Original to owner; 1 copy w/trophy; 1 copy retained by issuing command. If war trophy is firearm, DD 603 will also be prepared.

Theatre and inclusive period of service:
VIETNAM 1 Apr 69

Name of Owner (Last, First, MI): Bowers, Verne L.
SSN: 506-14-1019
Rank: BG

Organization: HQ, USARV (DCS P&A) APO 96375
Permanent Home Address: 3208 Circle Hill Road, Alexandria, Va. 22307

DESCRIPTION OF WAR TROPHY

Describe item (If firearm indicate Make, Model, Type, and Caliber):
NVA Mess kit

Serial number or identifying mark (if firearm include country of manufacture)
charred

Date: 26 Jul 69
Typed Name, Grade, Org of Authorizing Officer: ROGER G. TAFT, 1LT, Inf, Plans Officer

Station: Long Binh Post
Signature of Authorizing Officer:

DD Form 603-1

NVA mess kit, from the DMZ Run area. When I received it, it still had the soldiers ration of rice in it!

NVA Belt, taken off an NVA soldier killed in action by troops of 1st Brigade, 5th Infantry Division (Mech)

WAR TROPHY REGISTRATION/AUTHORIZATION

INSTRUCTIONS: This form will be prepared for all types of war trophies. Original to owner; 1 copy w/trophy; 1 copy retained by issuing command. If war trophy is firearm, DD 603 will also be prepared.

Theatre and inclusive period of service:
VIETNAM 30 Mar 69 - UNK

Name of Owner (Last, First, MI): Bowers, Verne L.
SSN: 506-14-1019
Rank: Brig Gen

Organization: DCS (P & A) HQ, USARV, APO 96375
Permanent Home Address: 3208 Circle Hill Rd, Alexandria, Va 22307

DESCRIPTION OF WAR TROPHY

Describe item (If firearm indicate Make, Model, Type, and Caliber):
NVA Pistol Belt w/metal buckle - star design

Serial number or identifying mark (if firearm include country of manufacture)
N/A NVN

Date: 19 July 1969
Typed Name, Grade, Org of Authorizing Officer: RAYMOND W. ANDERSON, SGM, Chief Opns SGT, LBP

Station: Long Binh Post
Signature of Authorizing Officer:

DD Form 603-1

George L. Wile was born in Worcester, Massachusetts on May 22, 1920 to a family of immigrants. His dad was from Nova Scotia Canada and his mom was of Finnish descent and came to the U.S. in the early 1900s. George was an extremely intelligent man and was unable to attend a formal college due to his parents financial situation like so many families in the early 1920, thus he decided to join the Army.

George entered the service through Fort Devens, Massachusetts and his basic training was in Texas. I remember stories of basic training and the snakes and scorpions. One of his bunkmates was stung while sleeping in his sleeping bag.
Dad (George) was shipped out in February 1943 to Camp Maxey, Texas according to a post card he sent to his parents on 7/11/1944. During basic training George learned to be a welder. This was a great job for him, since once he arrived in Europe his job was to repair tanks, guns, trucks, etc. away from the front lines. Many of the stories he told was of his travel with his unit throughout Italy, France, and the Czech Republic. Some of the post cards he sent home included one from Liebenstein town in the district of llm-Kreis, in Thuringia, Germany. Other locations were budetice u susice NA Sumave. This is located in the Czech Republic.
George loved to talk about his time in Europe and he never seemed upset or scared during his time there, although I always thought he had fonder memories then they in were actuality. One of these favorite stories was when he had to transport some German prisoners. They had fairly large truck and maybe 20 Germans and two hand guns for protection. They hid the bullets just in case they were overcome and this would help reduce their changes of getting shot. They had to drive at night with no lights in the truck and so many of the bridges were bombed out that the travel was extremely difficult driving down in to river beds, etc. Fortunately they made it safe. George also love to talk about living in bombed out churches, schools, etc. He never came face to face with active fighting but was stationed close enough to the front lines that they could hear the distant battles.
George was able to bring back a few things. He had three German Mauser rifles with bayonets, three dress daggers, a German flag, and a beautiful bellows camera; a Zeiss Ikon.
When George returned to the U.S. he used the skills he was trained for, for the remainder of his life and was a master welder. He even made a 10 foot ant out of metal spheres and had it in his back yard for years.

Of the souvenirs, just this item is left

George in France in 1944.

This is a very nice Nazi dagger that Dad brought back from WWII

Dad in uniform

In Europe his job was to repair tanks, guns, trucks, anything that needed fixing; but away from the front lines.

R.L. Clark, was a 1st Cavalry, 8th Regiment, 2nd Brigade veteran in WWII. He earned a Silver Star and 2 purple hearts for his service in the Pacific. He began in Australia, and then fought in New Guinea, Admiralties Islands, Leyte, Luzon, and then was stationed in occupied Tokyo, Japan. While in Tokyo the 1st Cav was disarming the Japanese and received a case of Type 94 Japanese pistols and other weapons. The Company Commander told the guys that they could take something home with them as a souvenir. He chose this Type 94 and brought it home. He was in service from 1943 to 1945.

I found out about this pistol from his daughter. One day she found out that I was a collector and asked if I would like to see the pistol her father brought back. She said that he kept it in a tool box in the garage, so it was rescued from the tool box and is now a great part of my collection.

Now it sits proudly in the collection of Shawn McKibben

Here is a K98k .22 conversion unit that was mailed home in its wooden box - the addresses are painted directly on the box. I got it at a show in the late 1980s, and the return address name was unusual so I looked it up in the phone book. Got a match in the same town. Talked with the guy and found out he was in an armor unit that trained in the southwest with new and sometimes secret equipment. They wouldn't let them go overseas because of their knowledge of the equipment they were testing. They finally got shipped over, arriving in the field just two weeks before the Battle of the Bulge. The guy was a watchmaker, and invented the illuminated reticle used in the tank gun scopes. He said he shipped a ton of stuff home during his time overseas, and that he had found the 22 unit under a bunk in a private room at the end of a barracks building in Aachen (obviously the room of the building NCO). When he got home in 1946 he sold all his souvenirs to start a business, and until he heard from me he had forgotten all about the .22 conversion unit. He was very thankful that I had called, and said, "You know, it really shouldn't matter to me, but I'm really happy that unit belongs to somebody who knows what it is and appreciates it."

Conversion unit with sender's name and address (Sgt. Lutter, a tanker in a heavy tank unit), sister's name and address, 'Passed U.S. Army Examiners' stamp, and a swatch of white paint with the examiner's signature on it. Stamped unit marking is for the German 1st Kompanie, Nachrichten [Signals] Abteilung Nr.6, Waffe Nr.2.

I got this PPK from my dad, who received it from his brother.

My uncle (Lieutenant Michael V. Lozano) was flying an observer plane as the pilot over Europe (don't know where) and was shot down. On the way down, he saw there were trucks and cars driving towards where he was going to crash land. He piled up the plane and pulled himself out before the Germans arrived. As his leg was injured, he improvised a crutch, lit the plane on fire, hoping the Germans would wait to see if the pilot had burned in the crash, stalling them in looking for him. He made his way to a forest and tried to make a plan on how to get back to the Allied lines.

After a couple of days he thought he was close to Allied lines so that night he took a road that he thought would take him back.

Walking past a couple of buildings, a car started and pulled out. As he could not move very fast with his injured leg, the lights of the vehicle were on him before he could get away. The car stopped and a gun fight ensued. The German was a Colonel and thought he had killed my uncle. As he approached, the major was stumbling and when the range was close, my uncle shot him.

My uncle said the major smelled like he had been drinking heavily. My uncle grabbed his pistol and then heard another vehicle coming, so moved as fast as he could to get away. Two days later my uncle got back to the Allied lines. The doctors wanted to amputate his leg, but he refused. He recovered well enough to go back into battle later on.

His injuries did 'bother' him the rest of his life, however, he said, that was the price of freedom when protecting your country. He did struggle some getting around in his later years. I never heard him complain or snivel once about his wounds (only that the doctors wanted to remove his leg).

Other than that, I have no idea how that PPK got from Europe to Southern California. I now have the PPK in my safe in Florida. When my dad is in his final resting place in Arlington (my Uncle Mike was eligible as well), I will more than likely donate all the 'family bring backs' to the Army Museum in their names as I have others from uncle Mike as well as rifles, swords and knives from the So. Pacific my dad brought back.

Additionally, I heard the above story more than a couple of times when growing up and the details were always the same.

Picked up by the WWII Veteran, he later in November 16th, 1945 turned it over to the Bemidji, MN Police Department, thinking they could use it. I was told that the Vets wife wanted him to get rid of it so he donated it to the PD. The Police kept it in storage for years and eventually sold or traded it to a class 3 dealer. From there it went onto the open market and I picked it up from a fellow collector.

1. Name of transferee ....City of Bemidji, (Police Department)....
(Trade name, if one is used)
2. Address of transferee ....City Hall.... (Street and number) ....Bemidji.... (City or town) ....Minnesota.... (State or country)
3. Kind of firearm ....German Machine Gun....
(State whether shotgun or rifle having barrel less than 18 inches in length, machine gun, submachine gun, muffler, silencer, etc.)
4. Serial No. ....4502m/44.... Model ....?.... Caliber ....9mm....
5. Other marks of identification ....MP 44....
6. Manufacturer of firearm ....Not known.... (State name and address)
7. Exemption from payment of the transfer tax imposed by section 3 (a) of the National Firearms Act, on the firearm described above, is based on the documents attached and herein listed, and other reasons as follows: ....Gun brought back as war trophy. Later decided to give the gun to the Bemidji Police Department for use in law enforcement....

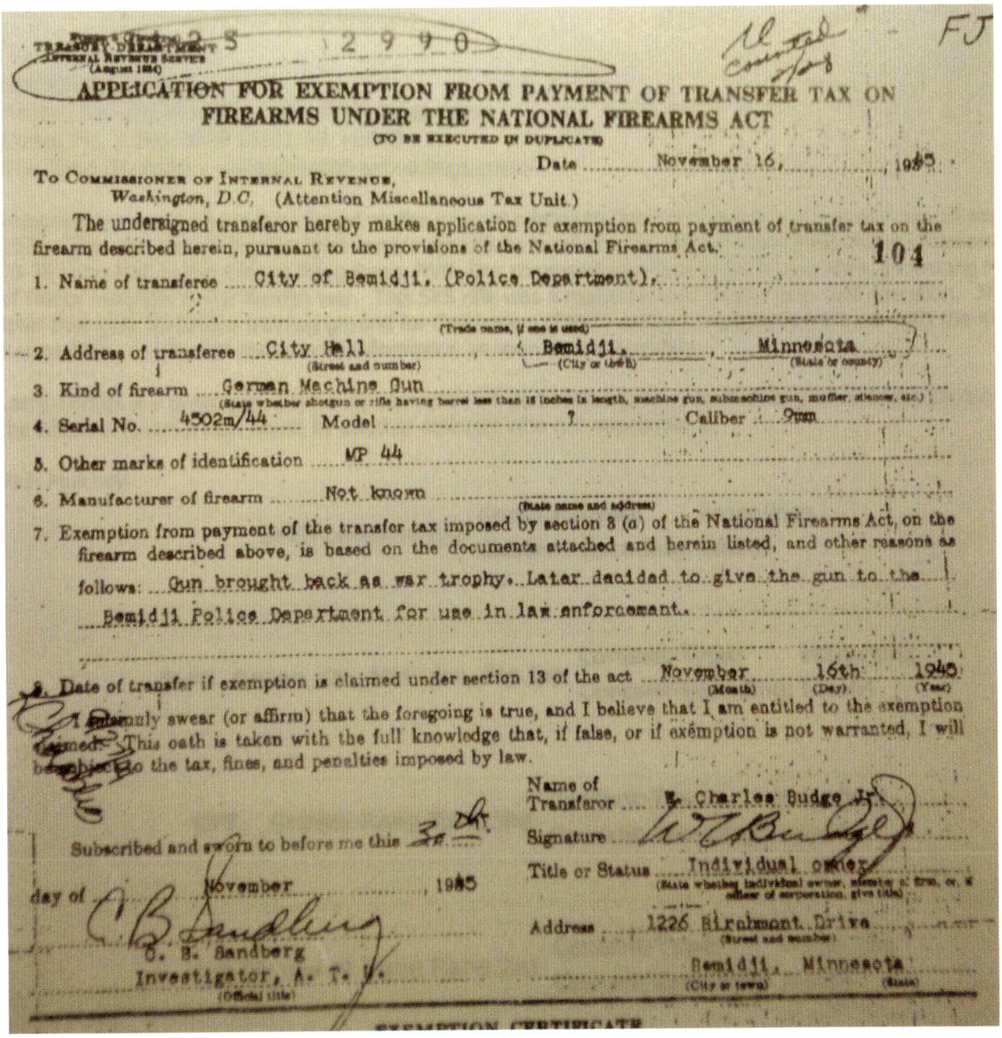

TREASURY DEPARTMENT
INTERNAL REVENUE SERVICE
(August 1934)

25 12990

FJ

APPLICATION FOR EXEMPTION FROM PAYMENT OF TRANSFER TAX ON FIREARMS UNDER THE NATIONAL FIREARMS ACT

(TO BE EXECUTED IN DUPLICATE)

Date November 16, 1945

TO COMMISSIONER OF INTERNAL REVENUE,
Washington, D.C. (Attention Miscellaneous Tax Unit.)

The undersigned transferor hereby makes application for exemption from payment of transfer tax on the firearm described herein, pursuant to the provisions of the National Firearms Act. **104**

1. Name of transferee City of Bemidji, (Police Department),
(Trade name, if one is used)

2. Address of transferee City Hall (Street and number) Bemidji, (City or town) Minnesota (State or county)

3. Kind of firearm German Machine Gun
(State whether shotgun or rifle having barrel less than 18 inches in length, machine gun, submachine gun, muffler, silencer, etc.)

4. Serial No. 4502m/44 Model ? Caliber 9mm

5. Other marks of identification MP 44

6. Manufacturer of firearm Not known
(State name and address)

7. Exemption from payment of the transfer tax imposed by section 3 (a) of the National Firearms Act, on the firearm described above, is based on the documents attached and herein listed, and other reasons as follows: Gun brought back as war trophy. Later decided to give the gun to the Bemidji Police Department for use in law enforcement.

8. Date of transfer if exemption is claimed under section 13 of the act November (Month) 16th (Day) 1945 (Year)

I solemnly swear (or affirm) that the foregoing is true, and I believe that I am entitled to the exemption claimed. This oath is taken with the full knowledge that, if false, or if exemption is not warranted, I will be subject to the tax, fines, and penalties imposed by law.

Name of Transferor W. Charles Budge Jr.

Subscribed and sworn to before me this 30th

Signature W. C. Budge

Title or Status Individual owner
(State whether individual owner, member of firm, or, if officer of corporation, give title)

day of November, 1945

C. B. Sandberg

C. B. Sandberg
Investigator, A. T. U.
(Official title)

Address 1226 Birchmont Drive
(Street and number)
Bemidji, Minnesota
(City or town) (State)

EXEMPTION CERTIFICATE

Dad was 20 when he signed up and 29 when he finally came out. He pretty well grew up on small amphibs and minelayers. Any credit goes to him. When you get into the details of what went on it is amazing what those GIs in both theaters accomplished with guts and American ingenuity. I believe we could never have won that war if we used the approach that the US is using today.

Attached is a rendition of the Amphibian Fifth Flotilla battle flag of Admiral Turner. It hung outside the mess hall with a sign "Kill Japs, Kill Japs, Kill more Japs." We were a little less Politically Correct in those days.

The group of guys in San Francisco

After Gaudalcanal Admiral Turner's island hopping campaign began. Munda Point and Vella Lavella were attacked by amphibious assault. The wrecked Japanese destroyer is near Munda Point. Munda was followed by a hard bloody fight on Bougainville. he Japanese rifle picture I believe is after Bouganville, and I think Dad has the 7.7 Arisaka, I believe the other is a T38. You can see they look a lot older than when they left San Francisco a year earlier.

Guadalcanal

Japanese rifles picked up and general clean up in Bougainville

The LCT proved to be a real workhorse for hauling things to the beaches but their slow speed on the open sea was just unacceptable. The secret weapon in the form of a large floating dry dock called a landing ship dock, LSD. Material can be found here http://www.navsource.org/archives/10/12/1205.htm

They would load 6 to 8 LCTs onto the LSD. Just drove them in and pumped the water. The LCTs were loaded with amphibious assault vehicles. They pumped the water out of the LSD dry dock and sailed away at almost 20 knots, 3 times the speed of the LCT. When the assault was staging, they flooded the dry dock and the LCTs took the amphib assaults into the first reef line, then opened their bow doors and the amphibs crawled on into the water over the reef and onto the beach. The attached photo shows an amphib coming back home after the landing zone was secure. Note the shell damage on the bow.

Dad said this one rammed up onto the bow to keep from sinking.

William Charles “Bill” Knechtel was born in May, 1926, in Baldwin Township, Pittsburgh, Pennsylvania. He served in the United States Army Air Corps from 1944 through 1946. He was not stationed overseas, but like any wartime veteran, he had his own ”bring backs” usually in the form of a good memory or story. But there were a few things “from the war” that he kept with him all of his life. While a student at Dormont High School, he was involved in various athletic teams and at thai time went by the knickname “Knute.” In October 1943, he enlisted in the USAAC and was on active duty in 1944. As the USAAC’s cadet program was being shut down, he was able to transfer from it into Kessler Field, Mississippi. There he was part of a volunteer group that built a temporary interlocking “matt” airstrip in three days. While he got orders for overseas, he caught tuberculosis ans was not allowed to go overseas.

One of his prize possessions is shown here in the picture taken by Kate Heinrich Anderson. It was given to me a month prior to his passing. The bracelet was originally just an id bracelet attached to steel link chain that holds a sterling silver, engraved “dog tag.” It was a Christmas gift from his parents that his sister Beverly engraved with a note that reads “From Mother & Dad 12-25-1943.” Later, she engraved his name and identification number on the other side. He had the Crew Wings added by a jeweler while in the service, he wore this during his service years and many times after.

While he never went overseas, he flew in B-17s, B-24s, and B-29s. He completed the requirements to wear the Gunner’s, Engineer’s, Observer’s, and Flight Crew’s wings. He crewed many flights beyond his required number, worked on the flight line, shot landings, and learned all he could about the planes and their engines at the bases he was stationed. .

One day, there was an exhibit of aircraft, some of which were “secret” and he began taking pictures, including an early picture of a P-80 with its pilot.

In addition, he was allowed to crew as an engineer, but there was a freeze on ranks and so they offered him that position as a Sergeant. He later was given the opportunity to become an officer, in exchange for an additional four year service. Coming from a family with artistic talents, he also helped to decorate the noses of planes and "inked" many jackets. His favorite character was Ms. Lace, which adorned his jackets. Unfortunately, those jackets have been lost to history. Anyone come across a Miss Lace flight jacket, contact the author please – especially if it is signed WCK or Knute!

He also was very interested in photography and took a series of pictures of airplanes that were on the tarmac at Clovis, New Mexico. A few of these are shown here. [Duel P51, Vrocket, one of the first P-80 prototypes]. As he explained these photos to me, he always carried a camera and while walking about the tarmac, one day, there was an exhibit of aircraft, some of which were "secret" and he began taking pictures, including an early picture of a P-80 with its pilot. This plane would crash in Sep 1946 in California. Later he learned that he was not to have been taking pictures.

During his time in the USAAC, he received a letter from a "buddy," Dorothy Lorenz, who was attending Capital University. They had first met as small children in a Sunday School class where she told him, someday I am going to marry you. They exchanged letters and he went to visit her via a "hop" to Columbus, Ohio to visit her at Capital College. Later he would fly with a Colonel as a crew chief on B-17s into Wright-Patterson. In 1946, they were married.

After his service, following a brief term with U.S. Army Map making working on maps of Vietnam in the late 1940s, he completed is Forestry degree from Penn State at Mt. Alto, and joined the US Forest Service from which he retired. He was a noted gunsmith and made many, many custom Mauser M98 hunting rifles. He also taught many the fine art of checkering and custom stock work. .He died on 14 Jun 2008, fittingly on Flag Day, still married to his sweet heart.

Gramps in 1944 with his wife.

At the time, top secret German V Rockets

Here is the story of how my dad got the Walther Model 4 with holster that I now own. Dad was in Co F, 385th Infantry Rgt, 76th Inf Division. His mother signed him in at 17 yrs old but he waited for his 18th to report for duty. He was a few days past 19 years old when he captured the gun. He was a rifleman, and before and after fighting played sax in the regimental dance band, later became a music teacher and professional musician, and taught music as a Junior College teacher.

Dad had told me many times about how he got this pistol which I thought was cool, but I was somewhat disappointed it was no Luger or P38. To me it was not a Nazi proofed gun and so what the heck. But I have changed my mind over the years for sure.

He wrote out his war time memoirs and yet for this particular story he had no recollection of the town, but it appears after one event and he is pictured wearing it in a photo; which I can definitely date, so I have a rough idea of where he was when he got the pistol but will likely never know the town. So, in his words I still have to set the stage.

“We were moving across Germany, and had occupied an Airfield at Altenburg-Nobitz on April 14, 15, or 16, 1945. From there moving east, towards the Mulde River and the Russians. Really a rat race at that time. The gun was captured between April 14th and April 26th 1945.

One day we came through a wooded area and around a curve in the road to a town. It was near the end of the day and we had been moving pretty fast. Although we had scouts along the wood, no enemy was flushed. We were pretty wary as we hadn’t had a good scrap in a couple of days--the enemy was on the run a lot, but would occasionally stand and fight. We cleared part of the town before dark and decided to just hold our ground until morning. We moved into houses and put the occupants into the basements. My platoon, along with First Platoon, occupied a house at the edge of town next to the road we had just traveled. Another house across the street was occupied by a machine gun platoon. We were covering the approach we had just taken; we suspected that the Germans had reoccupied the woods. A sentry was positioned in a 2nd story window facing the woods, and another was overlooking through the woodline the back yard on the town side. We also had a sentry in the driveway to the street.

At midnight I was awakened to relieve the guard in the driveway. As I was getting ready, there was a commotion at the back of the house. In the back yard was a German Lieutenant, accompanied by a private, who came to this particular house to have sex with the daughter. The German didn’t know that we were occupying the town, and he was taken completely by surprise by the sentry at the second floor window, who was covering him and hollered for help. The Lieutenant was brought up onto the back porch and I was asked to interrogate him. (Most of the interrogation for the company had been done by a guy in another platoon, but he was somewhere else, and I had by this time acquired phrase-book knowledge of German which served me pretty well). The German refused to surrender to an enlisted man; he demanded to be taken to an officer. I told him our commander was asleep and we weren’t going to wake him just for that, and I demanded that he hand over his pistol. He refused. I crowded him into a corner and demanded again. Refusing again, I butt-stroked him with my rifle, after which he quickly surrendered the weapon, a Walther .32 automatic. (In those days we were capturing many prisoners and acquiring lots of weapons. The rule came from the rear: each man was allowed to have one pistol for himself, and was instructed to pass the other pistols back to the rear for the cooks and others in the rear. Of course, we had our pick of the best, and lots of guys opted for Lugers and P-38s; I wanted something that was small and handy and concealable. Don't ask me why—that's what I was looking for. We were urged to give our pistols to the cooks to keep for us until the end of the war. If any of us were captured and found to have a German pistol on us, it was considered automatic justification for execution by the Krauts." Anyway, I still have that pistol, and I have fired it many times.

An hour later, there was another face to face encounter with the Germans where Dad fired some hip shots, and then threw a grenade; but they all got away.

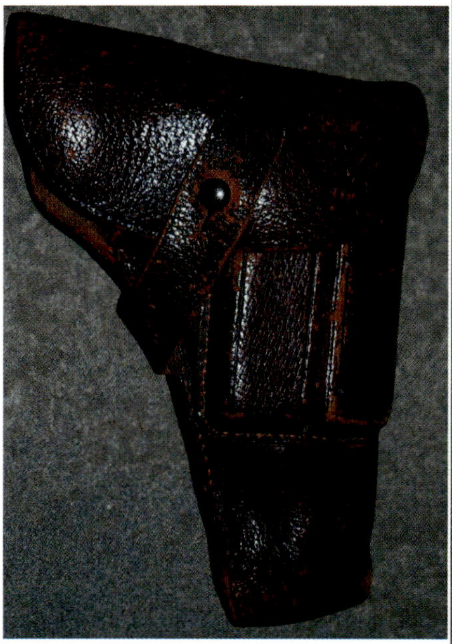

Later in Dad's memoirs, there is a story of a crashed airplane event and that story is basically that they were guarding bridges over the Mulde River at the end of April when a pair of Arado 96B trainers came in low. My dads friend had a perfect chance and emptied an entire 8 round clip into one of the Arados. All 8 were a hit, with two in the pilot. The plane crashed and the fuselage was torn from the wing. Everyone ran up to the plane and the pilot was dying as they pulled him out of the upside down fuselage. They laid him out on the field behind the wing and covered him with a blanket, you can see him in the photo. The lucky shot was not so lucky, he hunted rabbits in Montana so it was lots of skill involved. Most got a souvenir of the plane, my dad ended up taking pictures of everyone and then got his shot taken standing grinning in front of it. He is proudly wearing the pistol captured earlier and mentioned above.

It is a Walther Model 4, WWI production of the Second Variation with the later 6 raised slide grips. Serial 99xxx. The first gun I ever fired was that gun as a 12 year old, against metal plate targets at the local gun range. The holster is too big for the gun, has double magazine pockets on the front, but that is the one it was in when he got it.

Not much story, but good picture provenance!

**Major Phil C. Anderson and ARVN Officer
Military Assistance Command Vietnam, Advisory Team 64
(Phoenix Program)
Presentation of Captured Tokeroff Pistol (Type 54)**

More provenance on the Tokarev

REGISTRATION OF WAR TROPHY FIREARMS

The original of this registration form will be retained by the person authorized possession. This registration is not transferable.

1. THEATER RVN		
2. NAME OF OWNER (Last name, first name, middle initial) ANDERSON PHIL C.	3. SERVICE NO./SSAN 454-54-3785	4. GRADE O-4
5. ORGANIZATION MACV ADV TM 64 APO SF. 96215		
6. PERMANENT HOME ADDRESS (Street, City, State and ZIP Code) 414 MORRISSEY BLVD SAN CRUZ CALIF. 95060		
7. DESCRIPTION OF FIREARM		
a. MAKE CHICOM	b. TYPE (Rifle, pistol, shotgun, etc.) PISTOL	c. MODEL 54
d. SERIAL NUMBER 14087442	e. CALIBER 7.62mm	f. COUNTRY OF MANUFACTURE COMMUNIST CHINA
8. DATE 29 NOV 71	9. TYPED NAME, GRADE, AND ORGANIZATION OF EXECUTING OFFICER EDDIE L. COOPER, TSGT, USAF. NCOIC, Pass & ID Section 377th SEC. POL. SQ., APO 96307	
10. STATION TSN AB RVN	11. SIGNATURE OF EXECUTING OFFICER Eddie L Cooper	

Stamp: JOINT CUSTOMS GROUP Inspected and Passed SAIGON AREA DET

DD FORM 1 OCT 51 603

U.S. GOVERNMENT PRINTING OFFICE

WAR TROPHY REGISTRATION/AUTHORIZATION

INSTRUCTIONS: This form will be prepared for all types of war trophies. Original to owner; 1 copy with trophy; 1 copy retained by issuing command. If war trophy is a firearm, DD Form 603 will also be prepared.

Theater and inclusive period of service in overseas command REPUBLIC OF VIETNAM 4 DEC 70 2 DEC 71		
Name of Owner (Last, First, MI) ANDERSON PHIL C.	Service Number 454 54 3785	Grade/Rank O 4
Organization MACV ADV TM 64 APO SF 96215	Permanent Home Address 414 MORRISSEY BLVD SANTA CRUZ CALIF. 95060	
DESCRIPTION OF WAR TROPHY		
Describe Item (If firearm, indicate make, type - rifle/pistol) CHICOM PISTOL TYPE 54 7.62MM W/1EA MAGAZINE		
Serial Number or Identifying Mark (if firearm, include country of manufacture) 14087442 66 1967 COMMUNIST CHINA		
Date 28 NOV 71	Typed Name, Grade, Org of Authenticating Officer RICHARD F.TIBBETTS SFC,USA,CMEC	
Station APO SF 96307	Signature Richard F. Tibbetts	

Stamp: JOINT CUSTOMS GROUP Inspected and Passed SAIGON AREA DET

DD Form 603-1, 1 Jul 65

S.Sgt. Giret, Albert, P.
"The Rail-splitters"
US Army, Company I 335th Infantry Regiment, 84th Infantry Division
Rhineland, Ardennes, Central Europe

When we were kids we knew that my father was in the infantry during WWII. We also knew that he brought a "German Luger" back home. Once or twice a year we would hound him enough to get him to show it to us. He would order everybody to go upstairs, remove the pistol from its secret hiding place (in the basement workshop above a heating duct) and then call us back downstairs to admire his souvenir. Even as a kid I remember admiring the looks of that gun and being amazed at how the toggle operated. He didn't keep it out for very long. Maybe ten minutes or so. Then we were ordered upstairs again while he put it away.

My father told me that during the Battle of the Bulge he took the pistol, belt and holster off of a German Captain who was surrendering. They had been warned never to stand directly in front of an enemy soldier if you removed a pistol belt for fear of buckle guns. They were told to stand to the side or reach around from behind and pull the buckle loose. He told me the two things he remembered most about the event were the look of hatred the Captain gave him. "He was a German officer" Dad said. "He was a proud man and didn't want some punk touching him much less taking his gun." The second thing that stuck in his mind 20 some years later was the fact that standing next to the German Captain was a girl dressed in a German military uniform. When I pressed my father for further details he said "Some German officers had women on the front line' he said, and we used to call them "Fox hole warmers."

He wore the pistol belt for sometime. Finally, one night his outfit was sleeping indoors for a change. Wearing the holster made it difficult for my father to get comfortable so he took the belt off and hid it. The Luger went into his backpack. When he woke the next morning the belt was gone.

After the war ended things could get pretty dull in town. To pass some time they would go to a bridge in town and throw bottles or cans upstream and test their marksmanship by shooting at the cans as they floated past them. "If you didn't hit it you came pretty close" he said of the Luger. By that time he had three or four pistols. He was sent to school in England at some point. He couldn't take the pistols with him so he was told he had to turn them in before he went to England. He would be given a claim check and when he returned all he had to do was turn in the claim check and get his guns back. Needless to say he smuggled the Luger into England and when he returned and presented the claim check for the other pistols the reply was predictably "What guns"?

When my father passed away I fell heir to his 42 Code 1940. It's hard to appreciate everything he and countless other G.I.s went through to bring these trophies home. One can only hope that the stories they tell and the history written in them is never forgotten.

Here are some excerpts from letters my father wrote while in Europe, that mention the Luger and a Czech pistol that didn't make it back.

1) Letter dated Thursday, February 22, '45
" The reason the Filipino and I were together by ourselves is because we were flushing out a town and so (can't make it out) go into each house to clear it out of "nazty boys." We caught then eating. They had nice hot potatoes on the table. I ate one after we turned the Jerries over to some guards. Yum, yum. That's when the civilians started to kiss the hell out of us. We actually had to fight to get out of the house, no kiddin'. About sending the pistol home, there's a new order out forbidding the sending of firearms through the mail.

2) Letter dated Friday, March 30, 1945 Germany
"Now for that Jerrie pistol. I couldn't hide it because it is no small article. I had it in its holster and strapped around my waist. I wanted to get a good nights sleep for a change, so, for the first time in about three weeks, I took it from my waist and put it under the junk I was using for a pillow. I forgot about it in the morning, and went downstairs where we were oriented on taking the town of Golkrath, Germany. When I went back up, it was gone. By the way the taking of Golkrath was supposed to be written up in the 'Infantry Journal." My platoon was the first ones to go in there.

3) Letter dated Saturday, April 28, 1945 Germany
"Well I've got me a pistol, made in Czechoslovakia. How about that, huh kid?? On it, it says; Ceska Zbrojovka A.S. V Prage. I am not sure what that means (editor edited). But I intend to hang onto it.

"I've got a Jerrie pocket watch now, but it's not so hot. I had plenty of loot in my pockets the night those Jerries almost caught us. Loot was flying every which way. Everybody was throwing away Jerry watches, rings, bracelets, pens, pencils, coins, guns - anything that was Jerry property. So you can see how hopeless the situation looked. The Lord was with us again, as usual. I kept my pistol, but had it ready to heave when the time came.

Just a note ... My paternal grandparents were Slovak, hence the big deal about having a Czechoslovakian made pistol.

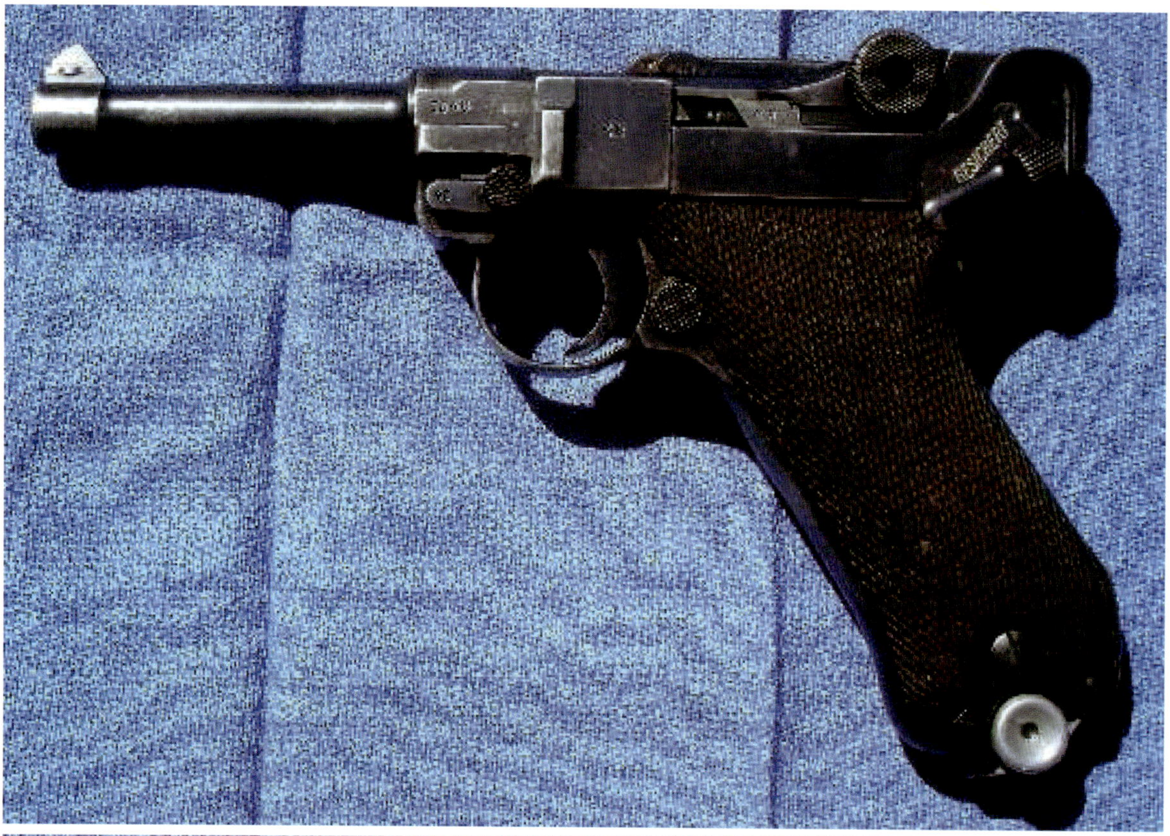

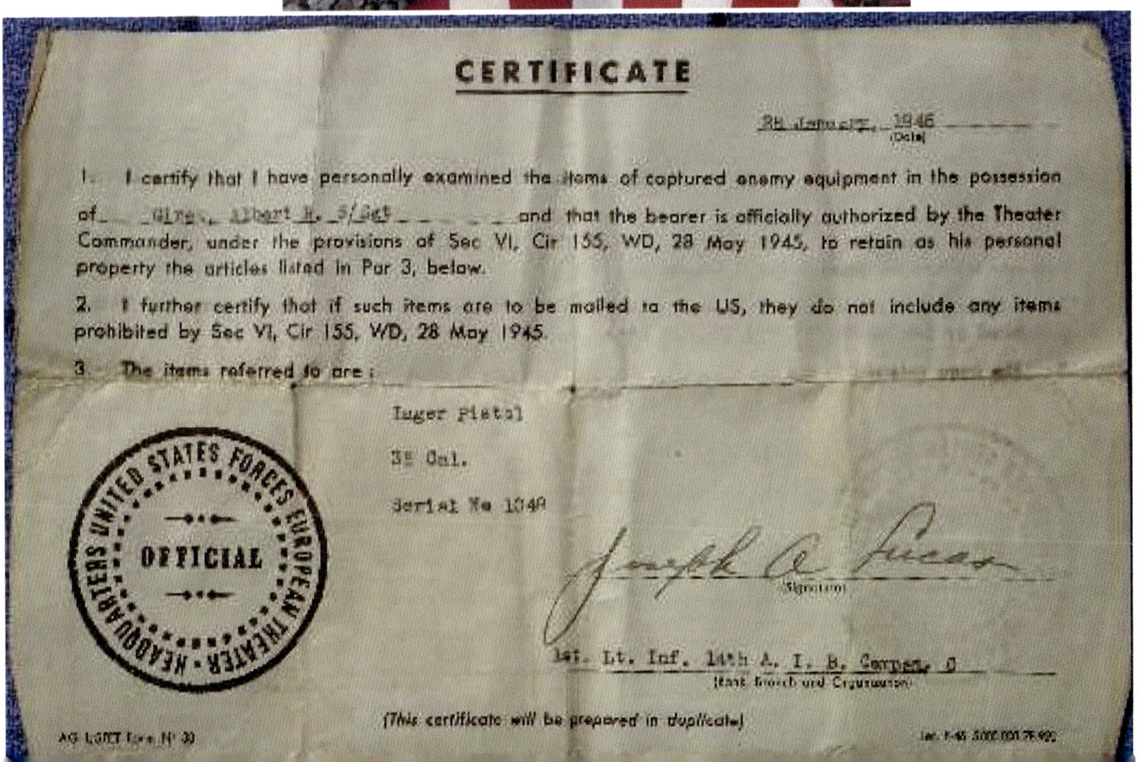

CERTIFICATE

28 January, 1946
(Date)

1. I certify that I have personally examined the items of captured enemy equipment in the possession of Giret, Albert H. S/Sgt and that the bearer is officially authorized by the Theater Commander, under the provisions of Sec VI, Cir 155, WD, 28 May 1945, to retain as his personal property the articles listed in Par 3, below.

2. I further certify that if such items are to be mailed to the US, they do not include any items prohibited by Sec VI, Cir 155, WD, 28 May 1945.

3. The items referred to are:

Luger Pistol

38 Cal.

Serial No 1048

HEADQUARTERS UNITED STATES FORCES EUROPEAN THEATER
OFFICIAL

Joseph A. Lucas
(Signature)

1st. Lt. Inf. 14th A. I. B. Comp. C
(Rank, Branch and Organization)

(This certificate will be prepared in duplicate)

AG USFET Form No 30

Some 20 years ago, or more, I used to run ads in the papers in the NY/NJ area for war relics. I got a call from a Veteran stating he had some Japanese items that he wished to part with. I met with him and asked what he had. He came out with the usual Type 99s, some Japanese ordinance, and an old battered NCO sword. In talking with him about his war experiences, he at first declined but slowly opened up. He told me a kind of weird story, which I took with a grain of salt - having heard my share of tales. He told me he had a Luger that he had picked up in the Pacific on one of the islands. I don't recall which island it was. I listened to the story, which was as follows:

He was walking along when he detected the unmistakable smell of death. Upon investigating, amongst a large set of rocks was the corpse of a Japanese soldier. Strangely, still in the soldiers hand was a German Luger pistol. The pistol had been laying on the ground with the soldier. The corpse was badly decomposed and the Veteran was reluctant to get too close. He described how he took a shoe lace and tied it around the barrel of the Luger, not wanting to touch the decomposed hand. As he tugged on the barrel, the Luger came free of the hand but was accompanied by the trigger finger. The Veteran took the Luger, along with the finger that he had wrapped in a rag. The Luger had a live round in the chamber and two left in the magazine. Upon listening to this story, I was saying to myself, "Oh, crap. Another tall tale."

The Veteran asked if I would be interested in buying the Luger. Up to this point, I did not even realize he still had the Luger. I indicated I would like to see it and he disappeared into the house and returned with the Luger. Upon examination, I noted the Luger was severely pitted on the complete left hand side from laying on the ground, as he had described. Then to my unbelievable surprise, he produced an old condiment jar containing the finger. I nearly fell over, never expecting the story to be true. It was a package deal and I believe I paid $60.00 for everything the Veteran had shown me on that day. Upon researching the Luger, it turned out to be a Dutch Luger. I believe the serial number was 125 and was dated 1928 on the frame. I'm not quite sure any more. I listed the Luger on GunBroker some 5 or 6 years back, along with part of the story. I believe this story also may have turned up on the Luger Forum, at that time (though I can't find it by searching.) The Luger generated huge interest on Gunbroker but everyone wanted the Luger, along with the finger, which I refused to sell. The truth is, from the beginning, I was not comfortable keeping the Japanese soldier's finger and felt that the finger should receive a proper burial, which I did in my back yard in NY, where it "rests in peace" to this day, hopefully with it's original owner. Believe it or not, this is a true Veteran story. I don't recall to whom I sold the Dutch Luger but regret selling it.

Ed, I wish I had documentation for you and the Veteran's name, along with the photos for your book - it would make one hell of a story to include. There you go, gentleman - that's the finger story. By the way, I had to see if the Luger still functioned properly as it was severely damaged on the left hand side. I ran three magazines through it with no problems.

US GI's with souvenirs they collected in the Pacific.

Although hard to see, there is a US soldier, advancing on suspected Japanese in the Pacific theatre, he is being covered by a GI with what appears to be a M1 Garand or M1 Carbine.

Action shots of pistols in war are very unusual to find.

As to the Belgian military Hi Power, there is nothing fancy. It came from Korea, from a member of the Belgian battalion. In April of 1951, one of the Belgian soldiers was injured and sent to a field hospital.

The gentleman I obtained it from was an orderly at the hospital. He had seen the pistol and liked it as it felt good in his hand. After a few days, when the Belgian soldier was to be sent home, he mentioned his like to him. The soldier told my gentleman to take it and keep it (with the comment that he would report it as being missing in the confusion of the battle).

Highly irregular as the Belgians kept good track of their weapons.

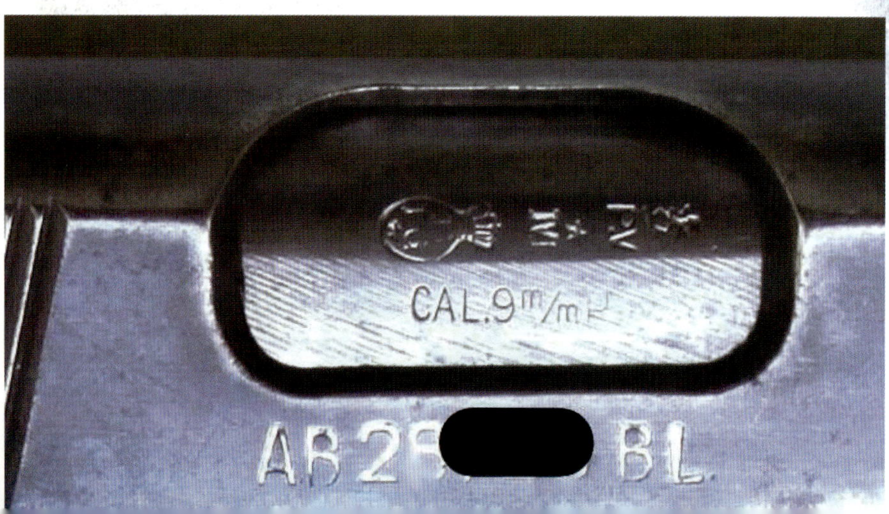

Nice British markings on the barrel

The other side of the Browning Hi Power

From the guy who sold me the pistol and met the Veteran.

I bought this pistol from a Veteran in 1978, can't remember his first name but his last name was Johnson and he was from Alma, Nebraska. He was a life time resident of this area and stated he was 35 years old when the war broke out so I guess he was born circa 1906 and he died about a year after I bought the pistol from him at age 72.

He said he was in southern Germany in April of 1945 and they had set up a camp near a stream. In the stream was a German body and after 3 days the captain ordered him to drag the body to the road so that someone from Graves Registration would take care of it.

He stated that the pistol was in a holster, and all he took was the pistol.

He didn't know how long the body was in the stream and said the soldier was an older man.

My Great Grandfathers name was Bayne Wesley Petty. He was attached to the 979th Field Artillery Battalion, 9th Army. He and his unit were somewhere near the Rhine River when he obtained this rifle. He came upon two Germans and he killed them both, in the process he shot the stock of the rifle with his M1 carbine. He took their rifles, and snapped some photos of them. He brought back a total of 6 rifles and two pistols, and this rifle is the only one he kept and we know where only one of his other souvenirs is.

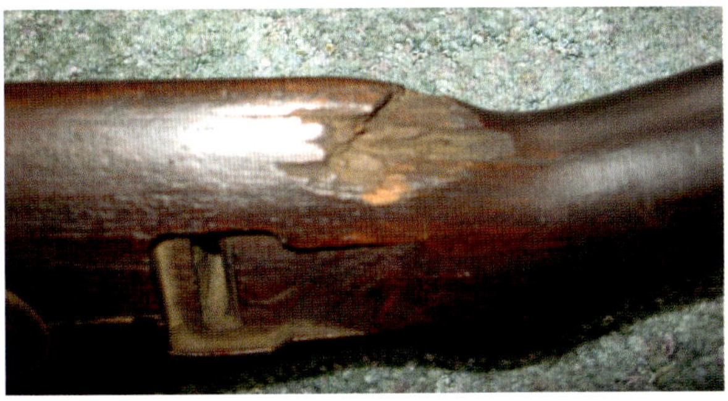

Here is the story on my dad, Donald C. Voigt. Born Nov. 18, 1919, he will be 91 this November. Dad was drafted in 1942 despite being in a "critical" profession of broadcast radio; given his training and background, he was sent to Signal school. Signal school turned into "Radar" school; a new and top secret specialty in 1942.

After school, Dad was sent to what was probably the best duty in the Army at the time, to set up and run radar units in the Caribbean, on Trinidad, a British Commonwealth island, just off the coast of Venezuela. In early 1943, three radar towers were constructed on Trinidad, by and later operated by the USA Signal Corps; their tasks were searching for enemy planes and tracking air traffic from the US to South America.

There was not much serious war activity around the island, only routine air flights, and the occasional possible sighting of a German submarine.

Though Dad stood guard duty with an M 1 carbine, and was issued a .45 pistol; the only "souvenir" he brought home was a US issue quartermaster's knife- a Cattaraugus 225Q.

He gave this to me at about age 10; I am amazed that I managed to play with it as a kid, carried it throughout Boy Scout days, and I still have it some 52 years later.

One of the three radar sites on Trinidad

The radar unit and tower

Dad just after he made Tech. Sergeant

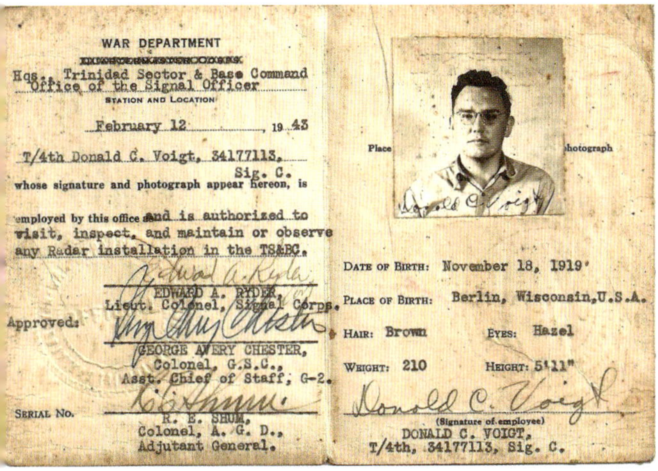

WAR DEPARTMENT

Hqs., Trinidad Sector & Base Command
Office of the Signal Officer
STATION AND LOCATION

February 12, 1943

T/4th Donald C. Voigt, 34177113,
Sig. C.
whose signature and photograph appear hereon, is

employed by this office and is authorized to visit, inspect, and maintain or observe any Radar installation in the TS&BC.

Edward A. Ryder
EDWARD A. RYDER,
Lieut. Colonel, Signal Corps.

Approved: George Avery Chester
GEORGE AVERY CHESTER,
Colonel, G.S.C.,
Asst. Chief of Staff, G-2.

SERIAL No. R. E. Shum
R. E. SHUM,
Colonel, A. G. D.,
Adjutant General.

Place hotograph

DATE OF BIRTH: November 18, 1919

PLACE OF BIRTH: Berlin, Wisconsin, U.S.A.

HAIR: Brown EYES: Hazel

WEIGHT: 210 HEIGHT: 5'11"

Donald C. Voigt
(Signature of employee)
DONALD C. VOIGT,
T/4th, 34177113, Sig. C.

His ID from Trinidad, and the manual ferry used to cross the river to one of the three radar sites on Trinidad

My dad was stationed at Scott Field, Illinois when the war ended and never got overseas. He was an Army Air-force cadet. He wrote my Grandfather that he purchased for 5 bucks a Nazi Hi Power that he would smuggle home to bring him a war trophy. I have the letter and the pistol (refinished at some point) and dad's uniform.

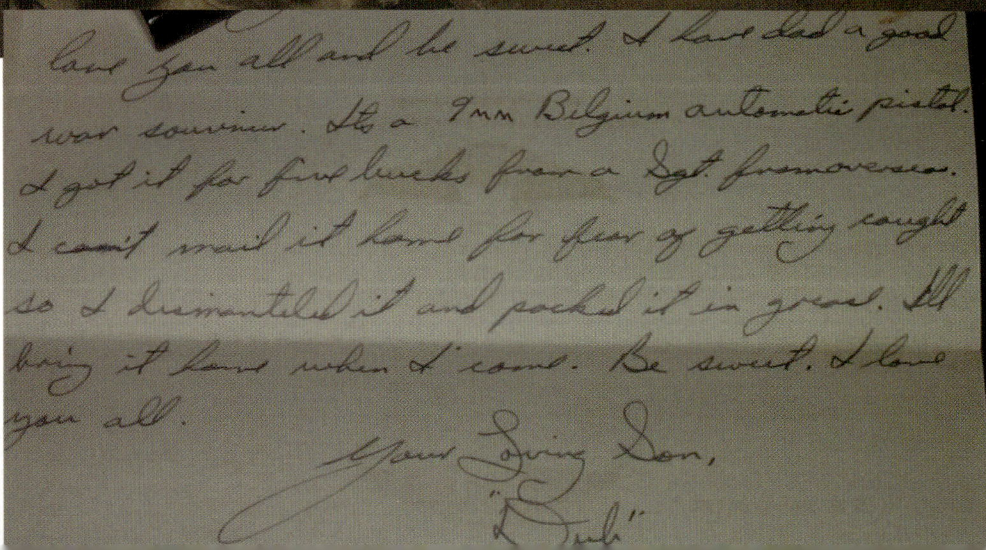

love you all and be sweet. I have had a good
war souvenir. Its a 9mm Belgium automatic pistol.
I got it for five bucks from a Sgt. from overseas.
I can't mail it home for fear of getting caught
so I dismantled it and packed it in grease. I'll
bring it home when I come. Be sweet. I love
you all.

Your Loving Son,
"Dick"

Pistol and belt buckle

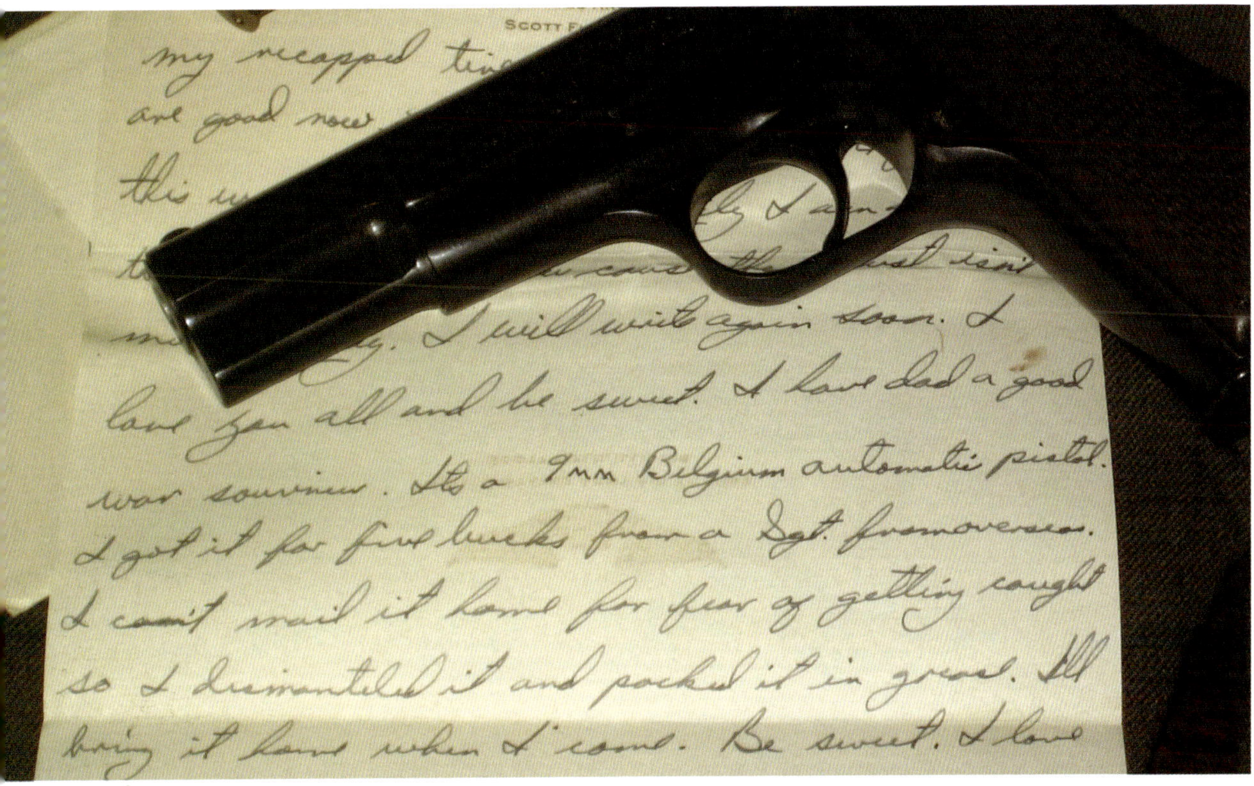

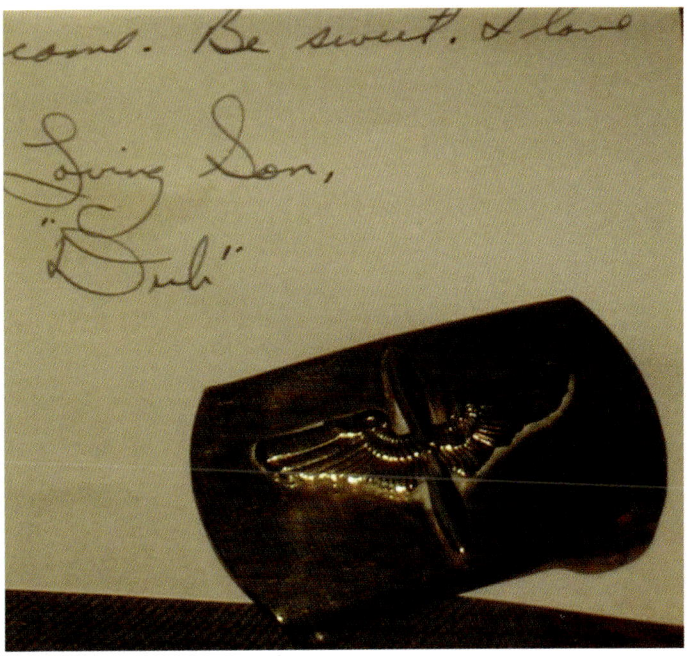

Shots of damaged equipment showing some of the ferocity of the Pacific fighting. The one shot is of A2 Douglas attack bombers left on Luzon in the 1942 surrender to the Japanese and photographed in 1945. Most of the rest the captions are all I know about them at this time.

Zekes on Munda—***Mitsubishi A6M Reisen Allied codename 'Zeke' or 'Zero Fighter'***

Japanese Navy Type 1 Attack Bomber on Munda—called by the US fighters, the Betty

Damaged aircraft

My uncle who brought this gun back is buried in Arlington Cemetery and had two tours in Vietnam. I think this would fit in the category "Why Bring it Back?"

From the missing parts, to the hole in the barrel, it still even has Vietnam dirt on it. I do not believe that it has ever been cleaned or touched since Vietnam. I certainly have no intention of messing with it.

It is a MAS 36, in rough shape, stock cut short for a small person. My uncle brought it back from Vietnam, and I do not have any paperwork. When I was 19 and living with him for a summer, I walked into his office. This MAS 36 was learning against a wall and I asked him where it came from. He said that his men (he was a Combat engineer officer) were plowing a road when they uncovered an arms cache. He let his men dig through what they found and no one wanted this rifle so he brought it home.

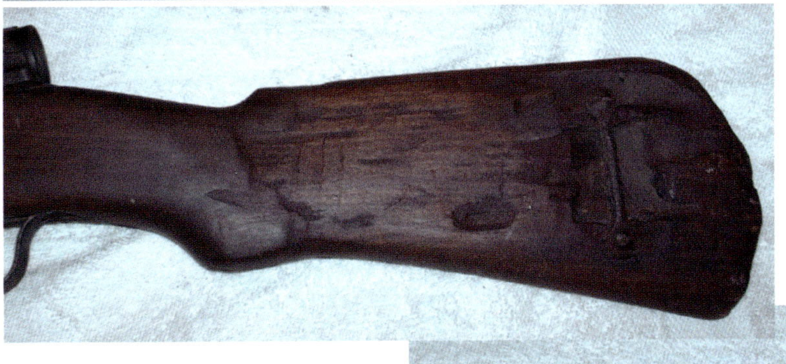

This pistol is named and highly engraved to an early Nazi German SS member named Thomas Jensen. In addition to the highly engraved pistol; there is a photo album of army life in the 1933-1935 time period; his soldier book with his handwriting and the serial numbered Mauser 1934 written in it.
Also included are his medals and assorted military documents, some copies, some original, pictures of his gravesite in Germany and both the soldier book and the album are made with a rich brown leather, a really fabulous and well documented item.

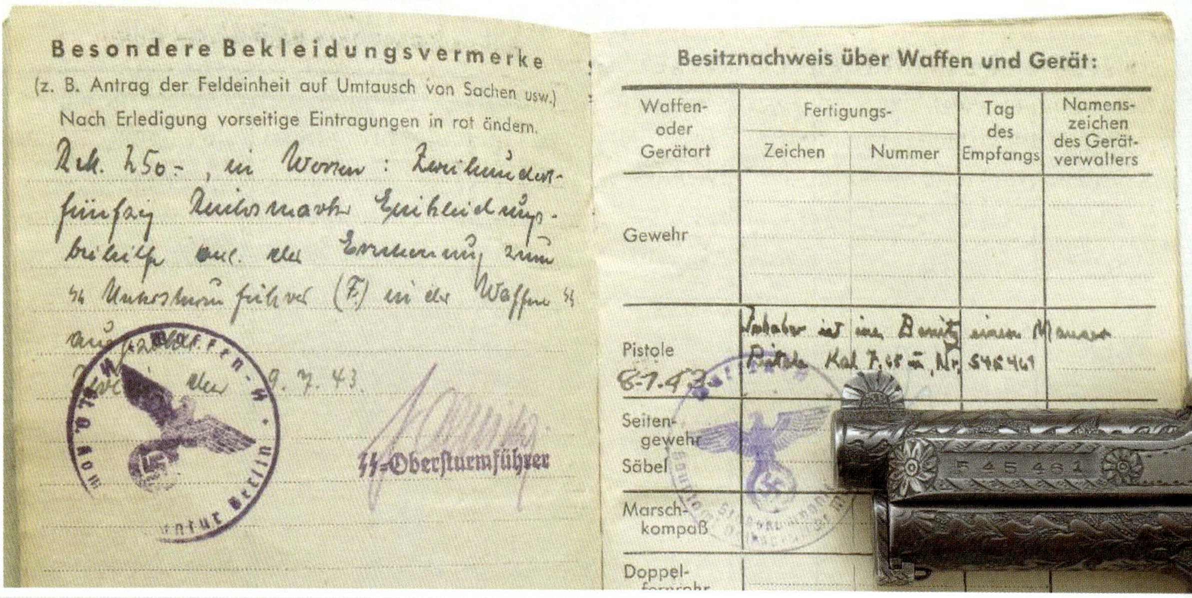

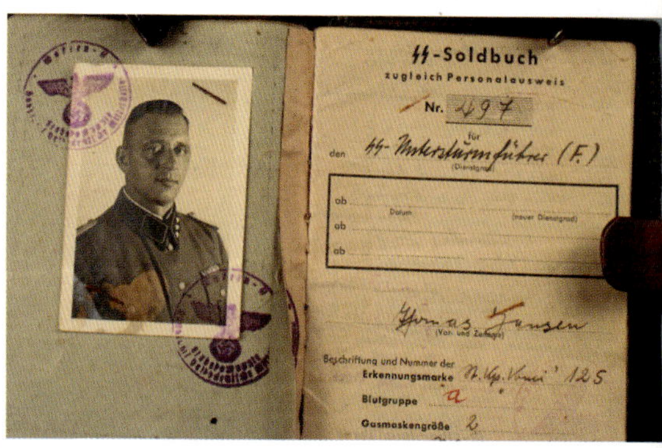

Even the rear sight is engraved!

Mike Morris, from whom I bought the gun, said he actually met with Jensen's family in Germany and visited his grave site.

Thomas joined the Nazi party pretty early in the movement, in the mid-30s or so. The upward arrow on the gun and book was the symbol for the Nazi SS officers training school. From the date in the soldiers book and since the gun was engraved with the schools symbol it is believed that he either had it done himself or the family had it done for him possibly as a graduation gift. The engraving was done outside the factory by a private engraver; not factory engraving. That probably meant that Thomas was from an upper middle class or wealthy German family.

Thomas Jensen

Named and highly engraved to an early SS member named Thomas Jensen.

Enclosed are pictures of a CZ27 with GI grips made from plexi glass. I believe this is an early model finish and bore are both good. This pistol came complete with holster and two correct magazines, and a TR winter cap and Hitler Youth armband.

This collection came from a Vets son who said his dad got the cap and CZ rig from two Germans he shot while laying telephone lines during the Battle of the Bulge. The Germans were in a motorcycle side car checking out a map ahead of their own lines. He used the cap under his helmet to help keep warm. The cap once had a Luftwaffe insignia which makes sense because the Germans used them as regular infantry during the Bulge. I'm also waiting for paper work regarding a Silver Star he won during this battle.

Markings on the top of the slide and the waffen on the side of the pistol.

Leonard Hrecho—French 1892

Back about 1999 I was approached by one of my fellow patrolman at work who told me that a Veteran's widow wanted to dispose of a gun and he had told her that the gun should be preserved, not destroyed and thought of me. He gave me the info and I went to the lady's house. We talked about her husband Lawrence who had been a Machinist in the Army during WWII. I later learned that he had been in the Normandy Invasion, Northern France, Rhineland and Central Europe and had earned five Bronze Stars.

Somewhere along the way he had picked up this French Revolver D' Ordonnance 1892 made by Saint-Etienne, serial number 30403 and brought it back in this pretty well used US 45 Holster. The widow wouldn't take my money for the gun so I left the money on the table in the kitchen when I left and thanked her assuring her that the gun would be kept with dignity in my collection.

The story should end there except for the fact that when her husband came home from the war, he stored the revolver in a suit case. It stayed in that suit case for the next fifty plus years and had gone on vacation with him and his wife. She told me that they had no children and traveled to Canada on a regular basis as they had a cottage there for many years. They also carried that suitcase with them as they travelled around the world to Europe and Spain in the 50s, 60s and 70s. Never an issue or never detected. This is one well traveled bring back.

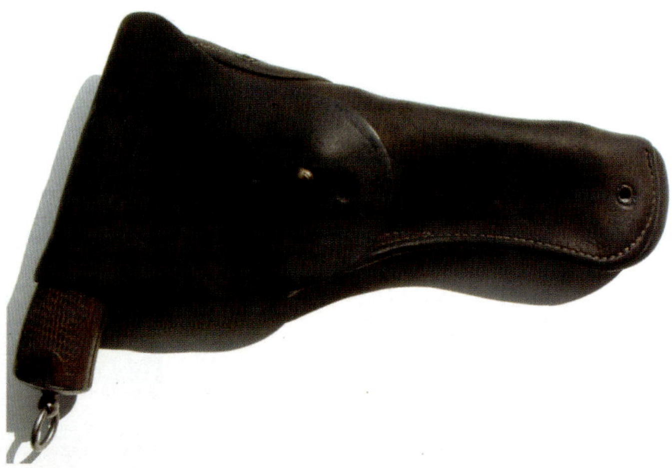

Laurence E. Magahey
Name

LAWRENCE E. MAGAHEY, T/5 36565708
Rank and Arm

2. As his Commanding Officer, I certify that the above named EM has by authority of this Theater Commander been authorized to retain the following items of captured material and has evidenced to me his right to possess the items of ~~British,~~ ~~Canadian, Russian,~~ FRENCH Government Material, ~~Allied Government Material.~~*
(Strike out words not applicable)
(itemize) REVOLVER, CAL. 32, #30403

[signature]
Signature

Capt [illegible] 9 Nov 45
Rank and Arm Date

3. The following customs declaration will be accomplished in all cases. In addition the customs declaration tag will be accomplished and affixed to the container.

CUSTOMS DECLARATION

I declare that all items herein consist of personal or household effects either taken abroad by me or acquired abroad for my personal use, except the following:
(here list or write "No Exceptions" as appropriate.)

NO EXCEPTIONS

Laurence E. Magahey
Signature

T/5 36565708
Rank and ASN

117
Tag No.

Type of Container DUFFEL BAG

(APPENDIX "C")

I bought the pouch directly from the Veteran with just a verbal story stating that he recovered it after a firefight with Viet Cong near Duc Hoa

Steve C—Papa Nambu pistol

I was given a TGE marked Papa Nambu from my dad over the weekend. I would say its good or better condition, although with a pitted lanyard ring and below it. It is Serial #4599, I took it apart and all numbers match, even the magazine and grip panels. My dad got it in Borneo, and said that they had been around and sharing things with the Australian soldiers and he didn't smoke so he traded an aussie carton of cigarettes for the papa!

Note: Borneo is the third largest island in the world and is located north of Australia.

Corporal Alfred Nielsen picked up a Nazi flag and had all his friends sign it. His daughter researched it and found all of these names in official records !

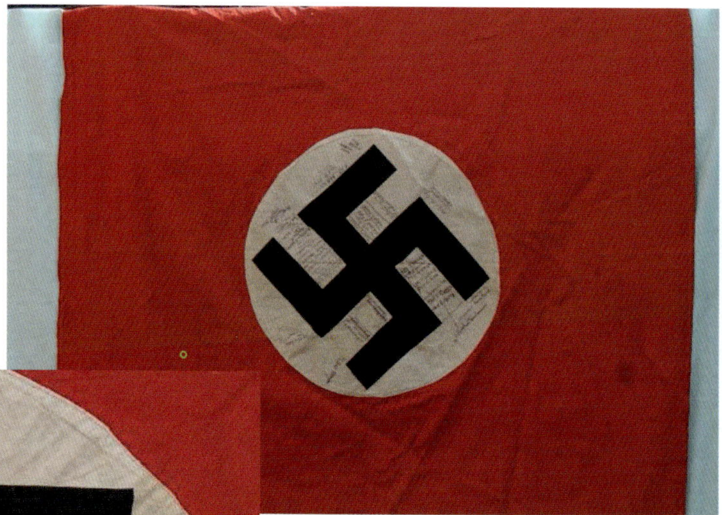

All Signatures on the flag are from soldiers who were in 740th Field Artillery, Headquarters Battery. I verified every signature against the unit's records at the National Archives, so if you have relatives and know the unit, it is possible to find out information!

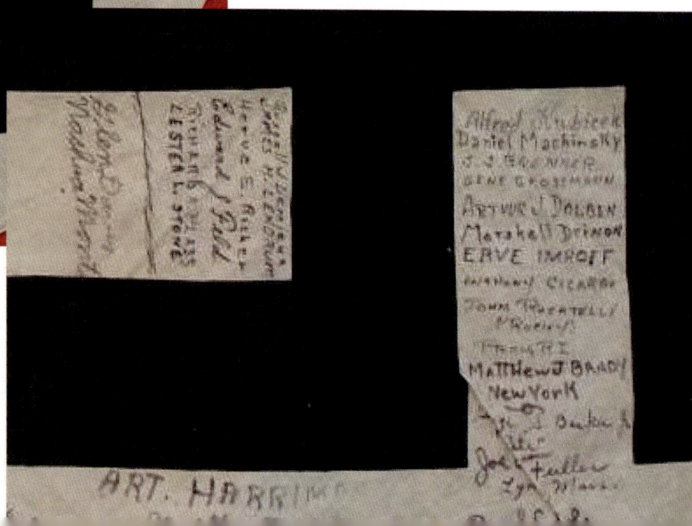

AK-47 Info

Background

As a senior Captain in November 1966 I joined the 1st Battalion 9th Marines, which had the nickname of "Walking Death." The battalion was in Okinawa and went aboard ship as the Special Landing Force (SLF). As such the battalion was utilized as a "put out the fire brigade." After an amphibious assault in the Mekong Delta in January, the battalion went ashore in the northern part of South Vietnam where all the in-country Marine forces were located. The battalion was in such heavy combat with so many casualties the nickname was changed to the "Walking Dead." I was assigned as the S-3A (Assistant Operations Officer) but after I was promoted to Major, I was moved up to S-3 (Operations Officer). During the heavy fighting against the North Vietnamese Army forces I got to know all of the Company Commanders very well and we became good friends. In the summer of 1967 the Commanding Officer of Company A presented me with an AK-47 that had been captured in the bloody battle for Con Thien, a combat outpost that was the battalion command post for several months.

Action

The following write-up of the battle is taken from the Marine Corps' *U.S. Marines in Vietnam -- Fighting the North Vietnamese 1967*:

"Lieutenant Colonel Richard J. "Spike" Schening's 1st Battalion, 9th Marines provided the security of the Con Thien perimeter. Two of the battalion's companies, A and B, on patrol a mile and a half northeast of Con Thien, had discovered a large NVA force, at least two battalions of the 90th NVA Regiment. The initial contact took place as the Marine companies were trying to link up. In this awkward situation, the North Vietnamese managed to punish each company separately. Additionally, the Communists hit the Marines with massed artillery fire from north of the DMZ (Demilitarized Zone). Casualties mounted on both sides as the battle intensified. Company B lost most of it's officers, the company commander, his FO (Artillery Forward Observer), the FAC (Forward Air Controller), and two platoon commanders were casualties.

Responding to this serious situation, Lieutenant Colonel Schening committed Company C, 1st Battalion, 9th Marines, reinforced by tanks.

The action intensified as the 3d Battalion, 9th Marines arrived from Dong Ha to help regain control of the situation. At this stage, III MAF (Marine Amphibious Force which usually has a division, wing, and a service support group. The Marine Corps has three in the active forces.) requested SLF Alpha. Similarly, the SLF Bravo, also at sea, went on alert. The mission assigned to SLF Alpha during Operation Buffalo, the name given the continuing engagement with the 90th NVA Regiment, was to tie in to the right flank of the 3d Battalion, 9th Marines and sweep the battle area to recover 1st Battalion, 9th Marines' casualties.

The original 0100 warning order for SLF Alpha preceded a regimental briefing at 0700 which set L-hour (the scheduled time for the first wave, usually LVT's to land on the beach, in this case time for helos to land ashore) at 1030. After much frenzied action on the part of the SLF, the launch took place on schedule. Throughout the late morning of 3 July, Lieutenant Colonel Kapetan's HMM-362 flew BLT 1/3 (battalion landing team which is an infantry battalion with artillery, tanks, helicopters, LVT's and a service support element) into the zone of action, supplemented by CH-46s from SLF Bravo's HMM-164. Unfortunately, the lead assault elements of the BLT landed 2,500 meters south of the designated landing zone, a mile southeast of Con Thien. Readjusting the battalion cost valuable time. The first physical contact with the right flank of the 3d Battalion, 9th Marines occurred at 1315. By the time all of BLT 1/3 had moved into the correct positions, however, the level of fighting had waned, and only Communist gunners remained active. The night of 3-4 July was relatively quiet, as was the next day. The 5th opened with an early morning enemy mortar attack against Companies A and C. Exchanges of artillery and mortar fire punctuated most of the day. The battalion continued its sweep operation. On the 6th, enemy rockets knocked out a Marine tank. The fighting around the disabled tank became so savage that the BLT did not reach the burnt-out hulk until 1000 on the 7th of July. The Marines recovered the bodies of the crew early the next morning.

At 1330, 8 July, the BLT received orders to make an immediate withdrawal. This was a challenging operation even under ideal circumstances. (LtCol. Peter A.) Wickwire's battalion faced a dangerous situation. There were five complications: it was daylight; the battalion was in contact; the withdrawal would occur over 600 meters of open ground; two of the attached tanks were crippled; and the battalion had

just received resupplies and had no organic transport to move the vital, though cumbersome, material. A sixth factor added to the BLT's problems. Just as the withdrawal began, the Communists opened up with small arms, mortars, and artillery. In spite of these difficulties, the battalion carried out this complex maneuver with only light casualties.

The next four days of Buffalo involved patrolling the Con Thien perimeter. Buffalo ended on the morning of 14 July. The Marines of BLT 1/3 had recovered the bodies of 11 fellow Marines from the 1st Battalion, 9th Marines, but their efforts were costly; the BLT's price was eight killed and 179 wounded. Operation Buffalo produced a total of 424 verified North Vietnamese dead for the BLT.

James D. Beans
Brigadier General, USMC (Ret.)

FORM 4467
(OCTOBER 1968)
Department of the Treasury
Internal Revenue Service

REGISTRATION OF CERTAIN FIREARMS DURING NOVEMBER 1968

(Chapter 53, Internal Revenue Code)

(Submit in duplicate for each firearm possessed. See instructions on reverse)

IMPORTANT This form cannot be accepted for registration of firearm except when received by Director during the time period November 2, 1968, through December 1, 1968.

To: Director, Alcohol and Tobacco Tax Division
Internal Revenue Service
Washington, D.C. 20224

I request registration of the firearm described below:

1. DATE: 19 NOV 1968

2. PLACE OF BUSINESS OR EMPLOYMENT: USMC

3. PLACE WHERE FIREARM IS USUALLY KEPT (See item 8 of Instructions): AT DUTY STATION: PRESENTLY - MCB QUANTICO VA.

4. NAME AND RETURN ADDRESS (Number and Street, City, State, ZIP Code): MAJOR J.D. BEANS, MOQ 396-D MCB, QUANTICO, VIRGINIA 22134

5. DATE OF BIRTH (Month, day, year): DEC. 21 1934

6. SOCIAL SECURITY NUMBER: 215-32-0780

7. DATE FIREARM ACQUIRED: 1967

8. KIND OF FIREARM (Machine gun, short barrel shotgun, etc. See instruction 5(A) on reverse): MACHINE GUN AK-47

9. SERIAL NUMBER (See Item 7 of Instructions): 10049350

10. LENGTH OF BARREL (Inches): 16 1/4"

11. OVERALL LENGTH (If a weapon made from a shotgun or a rifle or if in "Any Other Weapon" category):

12. CALIBER OR GAUGE: 7.62

13. MODEL: AK-47

14. NAME AND ADDRESS OF MANUFACTURER OF FIREARM: CHINESE COPY OF RUSSIAN AK-47 ASSUALT RIFLE

15. OTHER MARKS OF IDENTIFICATION: (66) 五六式

16. IF FIREARM IS UNSERVICEABLE EXPLAIN HOW IT WAS MADE UNSERVICEABLE (See Item 5(B) of Instructions)

Under the penalties of perjury I declare that this registration of firearm has been examined by me and to the best of my knowledge and belief, it is true, correct, and complete.

17. SIGNATURE: James D Beans

18. TITLE OR STATUS (State whether individual owner, member of firm, or if officer of corporation, give title. If registrant is other than a natural person, give name and home address of an executive officer thereof): MAJOR J.D. BEANS 072931 USMC INDIVIDUAL OWNER

FOR USE OF INTERNAL REVENUE SERVICE - APPLICANT WILL MAKE NO ENTRY IN THIS PART

19. ☑ Received for registration. This form when checked and signed below constitutes evidence of registration under the National Firearms Act of 1968 (Chapter 53 of Title 26, United States Code). Such registration does not permit the registrant to possess the firearm in violation of state or local law; however, see Item 6 of Instructions as to limitation on use of any information submitted herein.

20. DATE: JAN 22 1969

21. SIGNATURE, DIRECTOR, ALCOHOL AND TOBACCO TAX DIVISION: Harold ...

GPO : 1968 O - 338-050

FORM 4467 (10-68)

The AK and equipment came separately, but make a nice display

German SS pistol
serial #577780

My uncle 2Lt George A. Hendley was assigned to the 119th Infantry Regiment of the 30th Infantry Division serving in Europe during World War II. His duties were first as a S-2 staff officer then as a platoon leader. During the Battle of the Bulge, he was captured and held in La Gleize Belgium. When the massive allied counter-attack was lauched , he was killed in the evacuation of La Gleize. The gun, part of his personal effects, was sent to my grandfather. It has descended through my father to me.

Harvey Hendley

I have no doubt regarding the provenance. I suppose someone in theatre must have truly loved him and slipped it into the package of his personal belongings. I had tried to buy it from the sister with no success. Somehow my wife and older brother got together and got it for me as a surprise about 4 years ago. It was stored in the holster and very dirty, but the finish is beautiful. The lady is in her late 80s and lives a few miles from me here in South Carolina. He was a LT and a POW at Liege, unfortunately he was killed during the liberation of the prison. Contributed by Keith Smart

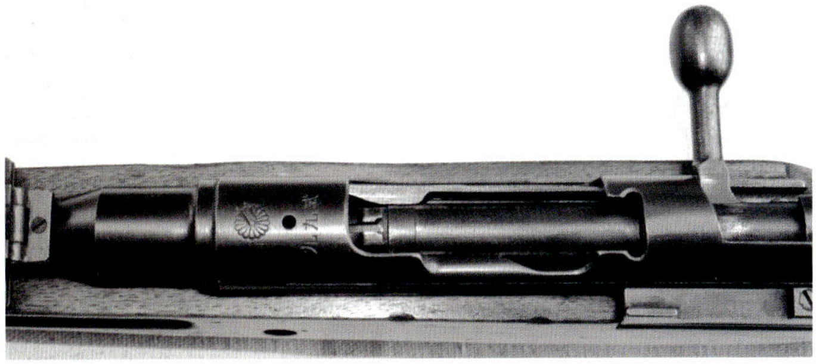

Here are photos of the 20th series Kokura Bob picked up at Tachikawa Air Base after the surrender. He told me that he heard there was a pile of rifles being picked up for souvenirs, so he jumped in his jeep and went to check it out.
He described an orderly pile of rifles and Light Machineguns stacked about 4 feet high and about 20 feet long. He picked up two rifles - one for him and one for his brother who was serving in the ETO. He said the rifle is exactly as he found it, stripped of the pod, rod, dust cover and AA wings. It has two strikes on the mum but is near mint otherwise.
Note that he states the rifle was taken "on the field of battle." He told me that permission was not actually given to acquire these rifles.

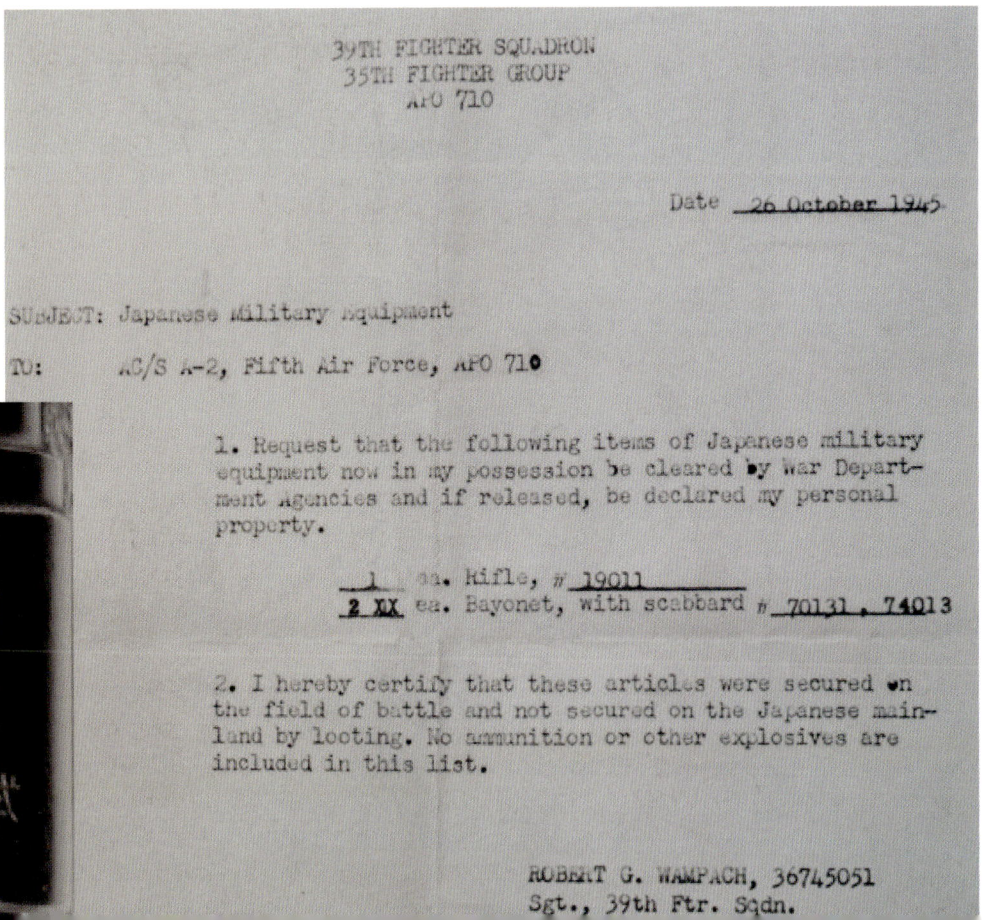

39TH FIGHTER SQUADRON
35TH FIGHTER GROUP
APO 710

Date 26 October 1945

SUBJECT: Japanese Military Equipment

TO: AC/S A-2, Fifth Air Force, APO 710

1. Request that the following items of Japanese military equipment now in my possession be cleared by War Department Agencies and if released, be declared my personal property.

1 ea. Rifle, # 19011
2 XX ea. Bayonet, with scabbard # 70131 , 74013

2. I hereby certify that these articles were secured on the field of battle and not secured on the Japanese mainland by looting. No ammunition or other explosives are included in this list.

ROBERT G. WAMPACH, 36745051
Sgt., 39th Ftr. Sqdn.

I have a short bring-back story of a small statue brought back after the war but it is a piece obtained during WWII - I'm not sure if it would be anything that you are looking for but thought I would share anyway. A year ago I was browsing local classifieds, when I came across a listing for a statue brought back from Germany. I called and the man I spoke to was the Veteran. who brought it back. He said it was a small figure of an explorer or something. I got directions and drove over to take a look. When I got there he brought out a 13" bronzed spelter statue of a Landsknecht Flag bearer. I told him I was interested but wanted the story of how he came about it. This is what he told me. While he was stationed in Augsburg 1955-1958 he befriended a German man by the name of Ernest May. Mr. May was crippled during the latter part of the war.

During the war he was in Munich and after a bombing he was searching for dead and injured in the debris. This is when in the rubble from a museum he spotted this statue. It was broken in about three pieces but he managed to find the pieces and hide them in his pants. He said he was afraid to get caught with it. He kept it put away until he told Mr. Haley Sowell (the Veteran. I was speaking to) about it. Mr. May gave this statue to Mr. Sowell still in pieces. He put it back together using an epoxy he obtained from another German friend. After he returned to the states Mr. Sowell who worked in an office for years kept this statue on his desk. He said he always liked it but he is older now and his interest has changed, so it needed a new home. I told Mr. Sowell I would gladly take it with the condition that he write his story for me which he did. I have tried to research it a little & have found that it was most likely made between 1910-1935 approx. The Landsknecht were used as a heroic image during Imperial & Third Reich times. Like I said not exactly a riveting capture but a post-war bring back of a different nature.

Jamie L. Stevens

The following statement is based on earlier telephone conversations and a personal interview that took place on Sept. 3, 2010. The items described in the following document are souvenirs that former Marine Cpl. Roy Kelley acquired while serving in the 2nd Marine Division in the Pacific theater. His tour included the islands of Saipan, Tinian, Okinawa and occupation duty in Japan.
Cpl. Kelly was trained as light machine gun crewman MOS 604, and as he stated during the interview he was "a very good machine gunner." From his position during the fighting on Okinawa, Cpl. Kelley observed the Japanese artillery fire that claimed the life of Commanding Lt. General S.B. Buckner.
Several uniform items include three Eagle, Anchor and Globe collar and hat insignia, one set of ID (dog) tags belonging to one Roy W. Kelley, SN# 856422 USMCR. One 2nd Marine Division patch. Personal awards include; Expert Rifleman shooting badge, ribbons that include Presidential Unit Citation with one battle star, WWII Victory, American Campaign, Asiatic Pacific Campaign.

The Japanese items include three coins and one Japanese Red Cross medal w/o ribbon. One coin is Chinese, the other period Japanese and a very old Japanese coin, a good luck piece? Also included is a Japanese Type 90 combat helmet. The helmet retains 90-95 % of the original paint, covered with a net, with one broken strand in front & back. The liner is intact with kanji on one pad, there are dark stains on one other pad, possibly blood. The chin strap is in poor shape, with dark stains, dingy and has been cut.

The origin of the items: the shooting badge is from Parris Island where Mr. Kelley's marksmanship landed him expert rifleman. The dog tags along with uniform insignia were issued at PI. The 2nd Marine Division patch was the shoulder patch for the 2nd MARDIV training at Camp Lejeune, NC. of which Mr. Kelley was a member. He remembers "just picking up the Japanese coins and stuff" while in the Pacific. The helmet came into his possession during the fighting on Okinawa. The Japanese would infiltrate the lines under the cover of darkness, one infiltrator, caught in the open, (under illumination) to the front of Cpl. Kelley's position was killed. Cpl. Kelley recovered the infiltrator's helmet the next morning. Mr. Kelley also brought back a Arisaka Type 99 rifle and bayonet that was separated from this grouping and sold elsewhere.

My Great-Uncle Riley Hughes was in the 701st Tank Battalion of the 4th Armored Division. He came home with his company book which included the map and song as well as the history of the unit in WWII. That book was given to me in 1986 and was the spark that started me collecting. The military picture is of my Uncle when he was in France. He was a tank driver and was wounded somewhere in Germany. The history of the 701st includes the liberation of the concentration camp at Gardelegen, Germany. Contributed by Stan Miller

"Tale of the 701" Sung to thetune of "Abraham"

We left the states, one cloudy day
Said goodbye, and sailed away
We're the701, We're the701

Arrived in England, the1st of May
Was there sometime, but not to stay
We're the701, We're the701

Crossed the channel with a skip and dance
And lande on the coast of France
We're the 701, We're the 701

France, Belgium, and Holland, we're on our way
Hit German soil on Christmas day
We're the701, We're the701

As we moved forward, they moved back
Here's the tale of our first attack
It won't be long, It wont be long

On February Twenty-Third
Thebiggest barrage, you ever heard
The attack was on, The attack was on

We crossed the Roer at the break of dawn
On the other side, we carried on
The attack was on, The attack was on
Chorus

In thisspearhead we became
Veterans of this war
Now we're glorious in our name
Rated tops by Corps

Now we're resting on the Rhine
Waiting fortheproper time
To drivethem on, To drivethem on

We'vefought hard through thick and thin
We'll drive on straight to Berlin
For when they slack, we'll go back
To the good ole U. S. A.

Words by
Cpl. John V. Edwards
U. S. Army

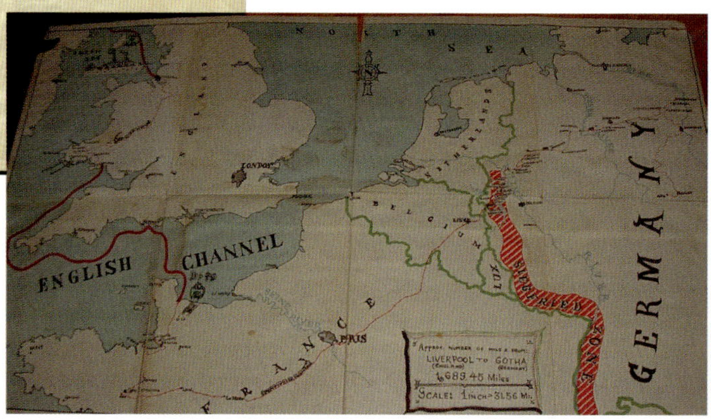

This German K98 was a gift from the Vietnamese gov't to the GI.

HEADQUARTERS
UNITED STATES MILITARY ASSISTANCE COMMAND, VIETNAM
DCSLOG/CLC Advisory Division, J463

MACJ463/MACV 2785

SUBJECT: Vietnamese War Trophy

1. Attached is a certificate authorizing you to obtain a war trophy through the Office of the Chief of Ordnance (OCO), RVNAF. Prior to release of the weapon OCO, RVNAF must issue a release order making reference to your name and the control number that appears in the upper right corner of your certificate. At the time the release order is cut, a decision is made as to which of the two ordnance collecting companies, located in the Saigon area, will release a weapon to you.

2. You must now contact, in person, the Chief of the Statistical Branch, OCO, RVNAF. This branch is located in Camp Phan Dinh Phung, Cholon, between the fish market and the Free World Forces Compound on Tran Quoc Toan Street. The map on the reverse of this letter will assist you in locating the Statistical Branch, OCO. The Statistical Branch will prepare the release documents and will tell you which depot will issue your weapon.

3. After you obtain the weapon you are responsible for processing it in accordance with MACV Directive 608-7.

4. This certificate has been signed by the Commander of the Central Logistics Command. He considers it a privilege to be able to give you a memento of your service to his country.

1 Incl
as

JOHN H. WHITE, JR.
MAJOR, USAF
Logistics Advisor DCSLOG/CLC, RVNAF

This German K98 was a gift from the Vietnamese gov't to the GI.

DEPARTMENT OF THE AIR FORCE
HEADQUARTERS 377TH COMBAT SUPPORT GROUP (PACAF)
APO SAN FRANCISCO 96307

REPLY TO ATTN OF: 377CSG/SP

SUBJECT: Transporting Firearms (TEMPORARY EXPORT PERMIT)

TO: Chief, Vietnamese Customs

This is to certify that LTC ARTHUR T. BIAGINI 185-12-6397
(RANK, NAME, SERIAL NUMBER)

has applied for an export permit on the following weapon:
1. 7.92mm Mauser rifle, mod 98, ser #9606, made in Germany.
////////////////////Last Item////////////////////

JOHN F. MURPHY JR, Lt Col, USAF
Chief, Security Police

CLEARED FOR CONUS
377th SEC. POL. SQ., APO 96307
15 JUN 1971

REGISTRATION OF WAR TROPHY FIREARMS

The original of this registration form will be retained by the person authorized possession. This registration is not transferable.

1. THEATER RVN		
2. NAME OF OWNER (Last name, first name, middle initial) BIAGINI, ARTHUR T.	3. SERVICE NO/SSAN 185-12-6397	4. GRADE LTC
5. ORGANIZATION HQ 7th AF APO 96307		
6. PERMANENT HOME ADDRESS (Street, City, State and ZIP Code) NEW STANTON, PA 15672 C/O Albert Biagini		
7. DESCRIPTION OF FIREARM		
a. MAKE MAUSER	b. TYPE (Rifle, pistol, shotgun, etc.) RIFLE	c. MODEL 98
d. SERIAL NUMBER 9606	e. CALIBER 7.92mm	f. COUNTRY OF MANUFACTURE GERMANY
8. DATE 15 Jun 71	9. TYPED NAME, GRADE, AND ORGANIZATION OF EXECUTING OFFICER JOSEPH SAPORITO, SMSGT, USAF Chief, Admin Security 377th SEC. POL. SQ., APO 96307	
10. STATION TSN AB RVN	11. SIGNATURE OF EXECUTING OFFICER	

DD FORM 1 OCT 51 603

WAR TROPHY REGISTRATION/AUTHORIZATION

INSTRUCTIONS: This form will be prepared for all types of war trophies. Original to owner; 1 copy with trophy; 1 copy retained by issuing command. If war trophy is a firearm, DD Form 603 will also be prepared.

Theater and inclusive period of service in overseas command REPUBLIC OF VIETNAM 3JUL70 to 3JUL71		
Name of Owner (Last, First, MI) BIAGINI, ARTHUR T.	Service Number 185 12 6397	Grade/Rank LTC 0-5
Organization Hq 7th AF (PACAF), APO SF 96307	Permanent Home Address NEW STANTON, PA. 15672 C/o Albert BIAGINI	
DESCRIPTION OF WAR TROPHY		
Describe Item (If firearm, indicate make, type - rifle/pistol MAUSER, Model 98 Rifle, 7.62mm w/equip		
Serial Number or Identifying Mark (if firearm, include country of manufacture) 9606 FED. REP. OF GERMANY		
Date 15 JUNE 1971	Typed Name, Grade, Org of Authenticating Officer HUGH H. WHITE, SFC, USA, CMSC	
Station APO SF 96307	Signature	

Joe bought this bring back and here is the story that he got with it:
My father Tech 5 Roscoe Tittle, was a member of the Army of Occupation in Japan in July 1945 assigned to Company D 720th MP Bn. As a member of the Military Police, he took this from a Japanese citizen.

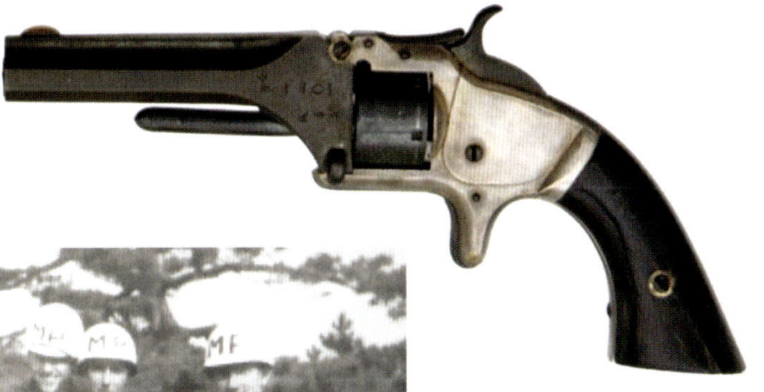

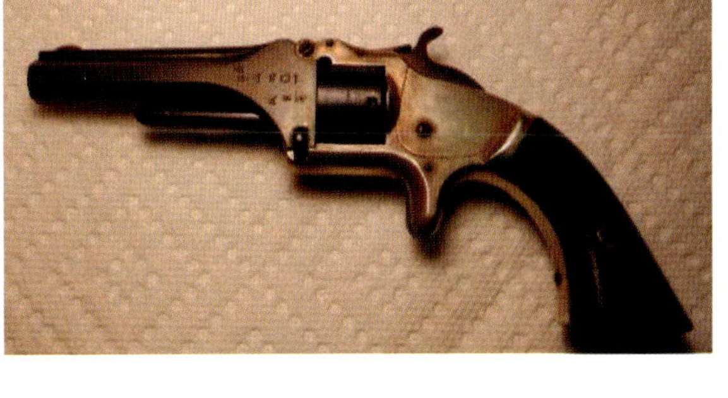

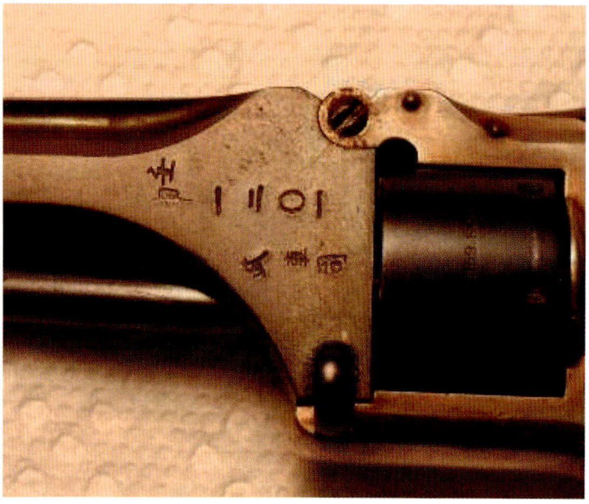

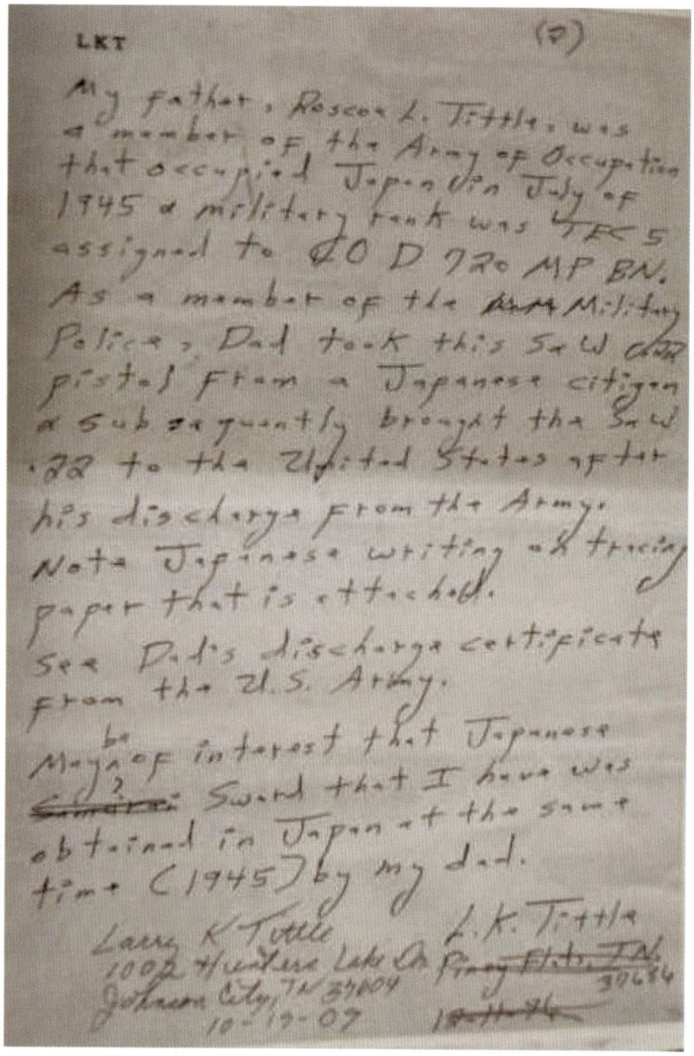

LKT (2)

My father, Roscoe L. Tittle, was a member of the Army of Occupation that occupied Japan in July of 1945 & military rank was TEC 5 assigned to CO D 720 MP BN. As a member of the ~~Army~~ Military Police, Dad took this S&W ~~[illegible]~~ pistol from a Japanese citizen & subsequently brought the S&W .22 to the United States after his discharge from the Army. Note Japanese writing on tracing paper that is attached.

See Dad's discharge certificate from the U.S. Army.

May be of interest that Japanese ~~[illegible]~~ ? Sword that I have was obtained in Japan at the same time (1945) by my dad.

Larry K Tittle L.K. Tittle
1002 Henderson Lake Dr ~~Piney Flats, TN~~ 37684
Johnson City, TN 37604 ~~[illegible]~~
10-19-07

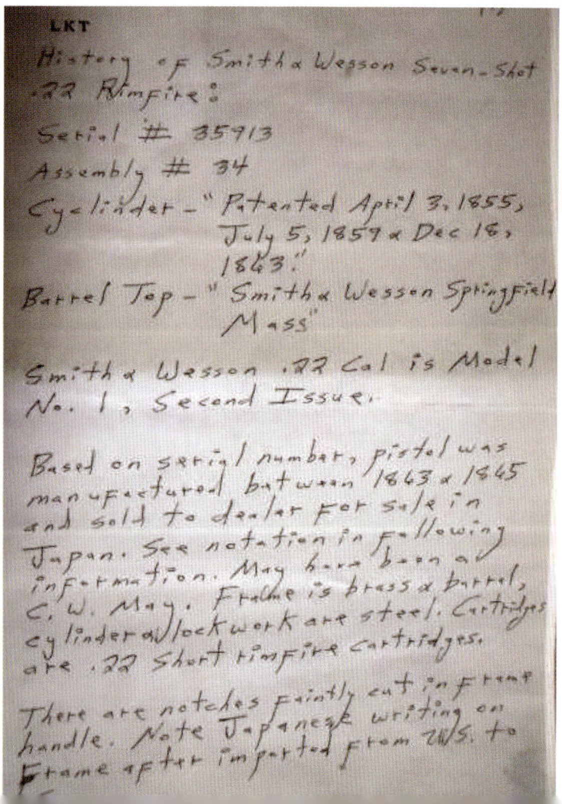

LKT

History of Smith & Wesson Seven-Shot .22 Rimfire:

Serial # 35913

Assembly # 34

Cylinder – "Patented April 3, 1855, July 5, 1859 & Dec 18, 1863."

Barrel Top – "Smith & Wesson Springfield Mass"

Smith & Wesson .22 Cal is Model No. 1, Second Issue.

Based on serial number, pistol was manufactured between 1863 & 1865 and sold to dealer for sale in Japan. See notation in following information. May have been a C.W. May. Frame is brass & barrel, cylinder and lockwork are steel. Cartridges are .22 short rimfire cartridges.

There are notches faintly cut in front handle. Note Japanese writing on frame after imported from U.S. to

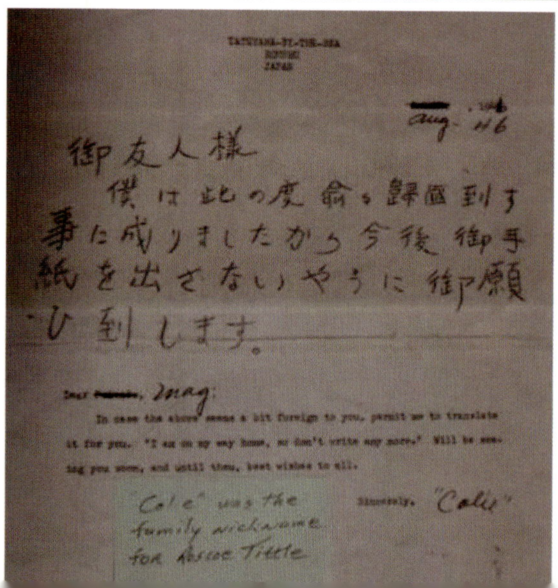

JAPAN

~~Aug.~~ Aug. '46

御友人様
僕は此の度愈々歸國到す事に成りましたから今後御手紙を出さないやうに御願ひ到します。

Dear ~~[illegible]~~, Mag:

In case the above seems a bit foreign to you, permit me to translate it for you. "I am on my way home, so don't write any more." Will be seeing you soon, and until then, best wishes to all.

Sincerely, "Cole"

"Cole" was the family nickname for Roscoe Tittle

One of my first Japanese rifles, with bayonet, I bought from Jack Tillison. He was on a Navy freighter, and he said that they were one of the first ships entering Tokyo harbor. He was an Ordnance man and was put in charge of all the black sailors, and it was their job when called to general quarters, to man the few 50 calibers and the 2 40 bofors. After anchoring the ship, the captain of the ship, sent Jack down to a near armory and told him to get a rifle and bayonet for every crewman on the ship. I think Jack said that he returned with 250 rifles and bayonets. The Capt. informed everyone that only the officers could have a rifle, and the enlisted men were given two quarts or fifths of booze. Jack said that some of the officers were pretty big drinkers so he was able to trade booze for rifles.

On the trip home some of the officers wanted Jack to test fire their rifle for them, because they had heard stories of Japanese rifles having trouble with head space and other problems, so Jack had to test fire their rifles. He said he would throw something in the water off the fan tail, as a target. Then he would point the rifle towards the water, and turn away from the rifle, turning his head and close his eyes, and pulling the trigger. He then took the rifle back to it's owner and reported that the rifle was dead on. Below is a picture of the rifle and bayonet. The rifle is an I type and the bayonet has a hint white painted Kanji on the edge of the tang. So I think that it helps validate his story of the pieces coming from a Naval arsenal.

Corporal Alfred Nielsen did something very cool. He collected European money and wrote where he had been on them during WWII

Aug. 10-"44".
St. Malo - Brêst - Crozon Pnn. - (Rennes - Le Mans
Chartres - Paris - Sedan) Oct. 2-"44"
100
181
BANQUE DE FRANCE
P.13022
100
CENT
FRANCS
DO. 1=8=1940. DO.
325539181
LE CAISSIER GÉNÉRAL,
LE SECRÉTAIRE GÉNÉRAL,
P.13022
181

"A. of O." Regensburg - Furth - Frankfurt/M. -
Urb
Grosslengelfeld
Prum
Bollendorf
Enzen
Bitburg
Kaiserich
Cerden
Simmeren
Kreuznach
Alzey
Worms
(Crossed Rhine Mar 29)
Darmstadt
Frankfort/M
Hanau
Fulde
Meiningen
Coburg
Bayreuth - Nabburg - Cham - Passau - Linz (Aus.) "May 8"
Reichsbanknote
RBD
100000
Mark
Hunderttausend Mark

THE PRESIDENTIAL UNIT CITATION (ARMY) FOR EXTRAORDINARY HEROISM TO THE 761st TANK BATTALION, UNITED STATES ARMY

The 761st Tank Battalion distinguished itself by extraordinary gallantry, courage, professionalism and high esprit de corps displayed in the accomplishment of unusually difficult and hazardous operations in the European Theater of Operations from 31 October 1944 to 6 May 1945. During 183 days in combat, elements of the 761st - the first United States Army tank battalion committed to battle comprised of black soldiers - were responsible for inflicting thousands of enemy casualties and for capturing, destroying, or aiding in the liberation of more than 30 major towns, 4 airfields, 3 ammunition supply dumps, 461 wheeled vehicles, 34 tanks, 113 large guns, 1 radio station, and numerous individual and crew- served weapons. This was accomplished while enduring an overall casualty rate approaching 50 percent, the loss of 71 tanks, and in spite of extremely adverse weather conditions, very difficult terrain not suited to armor operations, heavily fortified enemy positions and units, and extreme shortages of replacement personnel and equipment. The accomplishments are outstanding examples of the indomitable spirit and heroism displayed by the tank crews of the 761st. In one of the first major combat actions of the 761st, in the vicinity of Vic-sur-Seille and Morville-les- Vic, France, the battalion faced a reinforced enemy division. Despite the overwhelming superiority of enemy forces, elements of the battalion initiated a furious and persistent attack which caused defending enemy elements to withdraw. While pursuing the enemy, tanks of the 761st were immobilized before an anti-tank ditch. Savage fire from enemy bazooka and rocket launcher teams, positioned 50 yards beyond the ditch, disabled many of the vehicles. Crewmen dismounted the disabled tanks, resulted in the elimination of many of the positions and virtually destroyed two enemy companies while permitting the escape of other tanks and crews and eventual completion of the mission. From 5 January 1945 to 9 January 1945, the 761st engaged the 15th Panzer Division in the vicinity of Tillet, Belgium. Suffering severe casualties and damage to their tanks, the 761st attacked and counter-attacked throughout the five-day period against a numerically superior force in both personnel and equipment , and on 9 January 1945 the men of the 761st routed the enemy from Tillet and captured the town. This action was significant in that the enemy was prevented from further supply of its forces encircling Bastogne, and the United States troops there, because of the closing of the Brussels-Bastogne highway by the men of the 761st. One of the most significant accomplishments of the 761st began 20 March 1945 when, acting as the armor spearhead, the unit broke through the Siegfried Line into the Rhine plain, allowing units of the 4th Armored Division to move through to the Rhine River. During the period 20 March 1945 to 23 March 1945 the battalion, after operating far in advance of friendly artillery, encountered the fiercest of enemy resistance in the most heavily defended area of the war theater. Throughout the 72-hour period of the attack, elements of the 761st assaulted and destroyed enemy fortifications with a speed and intensity that enabled the capture or destruction of 7 Siegfried towns, 31 pill-boxes, 49 machine gun emplacements, 61 anti-tank guns, 451 vehicles, 11 ammunition trucks, 4 self-propelled guns, one 170mm artillery piece, 200 horses, and one ammunition dump. Enemy casualties totaled over 4,100 and of those captured it was determined that the 761st in its Siegfried Line attack had faced elements of 14 different German divisions. The accomplishments of the 761st in the Siegfried area were truly magnificent as the successful crossing of the Rhine River into Germany was totally dependent upon the accomplishment of their mission. The men of the 761st Tank Battalion, while serving as a separate battalion with the 26th, 71st, 79th, 87th, 95th and 103rd Infantry Divisions, the 17th Airborne Division, and 3d, 7th, and 9th Armies in 183 continuous days in battle, fought major engagements in six European countries, participated in four major allied campaigns, and on 6 May 1945, as the easternmost American soldiers in Austria, ended their combat missions by joining with the First Ukrainian Army (Russian) at the Enn River, Steyr, Austria. Throughout this period of combat, the courageous and professional actions of the members of the "Black Panther" battalion, coupled with their indomitable fighting spirit and devotion to duty, reflect great credit on the 761st Tank Battalion, the United States Army, and this Nation.

Signed Jimmy Carter, President of the United States of America

Here's an M1940 helmet that I found at a local garage sale a couple of years ago. I thought you guys might appreciate it. According to the guy running the garage sale its an old German helmet and "there's some black soldier painted on top of it, there might be some history behind that." So, meet 1st SGT. Robert L. Linzy. He served 18 and a half years in the Army, he was lucky and survived the war (his battalion suffered a 50% casualty rate).

The 761st Tank Battalion (AKA. Black Panthers Tank Battalion) was the first African American tank unit to serve in the U.S. Army. They were intended to be an experiment similar to the better known Tuskegee Airmen.

Although General Patton doubted that blacks could operate a complex machine like a Sherman tank in combat, he decided to keep the 761st at the front of his advance through Europe. The 761st went 183 consecutive days with no relief or rotation from the front, at a time when the average U.S. tank unit had a 17 day life expectancy at the front. Their combat record is very impressive. If you are interested I would suggest doing a search on the 761st as there is a lot of info on them. Jackie Robinson was in this unit until he was discharged for refusing an order from an army bus driver to sit in the back of the bus (Robinson was an Officer), and one member was awarded the Medal Of Honor for his actions in the Bulge. They were also part of Task Force Rhine and were one of the first units to meet up with the Red Army at the river Steyr.

Here is a Photo of Sgt. Linzy on his Sherman while he was part of the occupational force after the war (he stayed until some time in 1946). I am guessing this is when he had the time to paint captured helmets. There is another known helmet similar to mine that he painted and mailed home as a gift to his sister, it is in a museum in Geneva NY. I would imagine that there may be more of his helmets out there?

The Veteran states he took this from a German officer and that he had it till selling it to my friend in 2003. The Veteran was in the 405th Inf Reg, 102nd Div. and its his photo you see on the next page. The gun is an all matched AC 44 including the holster dated 44 as well. The letter states spare magazine, but my friend lent it out to some buddy never to be returned. The boots in the back ground are the ones worn by the Veteran in the photo; he has his name in ink in them.

The boots in the back ground are the ones worn by the Veteran in the photo; he has his name in ink in them.

This SKS came out of the Cu Chi area of Vietnam and was brought home by a Veteran.

A little history for you.....I had two tours in Viet Nam....May 67-May 68 and Oct 70-May 72. Both tours were with the 175th Radio Research Company (later the 175th Radio Research Field Station). This was an intel unit under the Army Security Agency (later Inscom). The weapon came from a weapons cache in the Cu Chi area just to the north of our location at Bien Hoa Air Base. One of my buds was able to obtain two (one for him and one for me). Got it in the early part of 1971 but cant remember the exact date. My bud was an advisor with the ARVNs. I supplied him with items he couldnt get and he kept me supplied with stuff the ARVNs picked up or captured. At one time or another I had a Thompson machine gun, a browning pistol, and a pair of the soviet pistols. The SKS was the only item I brought home. bob

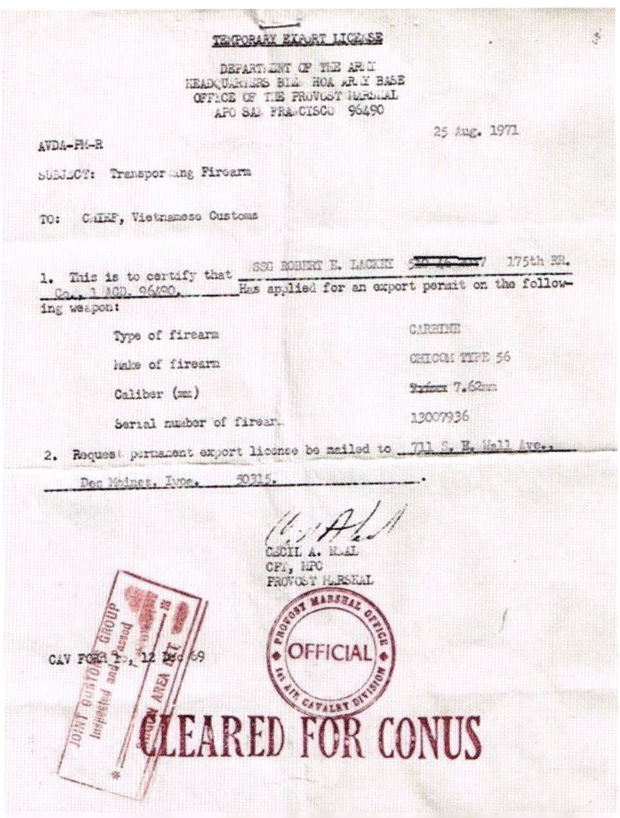

TEMPORARY EXPORT LICENSE

DEPARTMENT OF THE ARMY
HEADQUARTERS BIEN HOA ARMY BASE
OFFICE OF THE PROVOST MARSHAL
APO SAN FRANCISCO 96490

25 Aug. 1971

AVDA-PM-R

SUBJECT: Transporting Firearm

TO: CHIEF, Vietnamese Customs

1. This is to certify that SSG ROBERT E. LACKEY [redacted] 175th RR. Co., 1 ACD, 96490, Has applied for an export permit on the following weapon:

Type of firearm	CARBINE
Make of firearm	CHICOM TYPE 56
Caliber (mm)	7.62mm
Serial number of firearm	13007936

2. Request permanent export license be mailed to 711 S. E. Hall Ave., Des Moines, Iowa. 50315.

CECIL A. HEAL
CPT, MPC
PROVOST MARSHAL

CAV FORM 1, 12 Dec 69

JOINT CUSTOMS GROUP
Inspected and Passed

PROVOST MARSHAL OFFICE
OFFICIAL
1st AIR CAVALRY DIVISION

CLEARED FOR CONUS

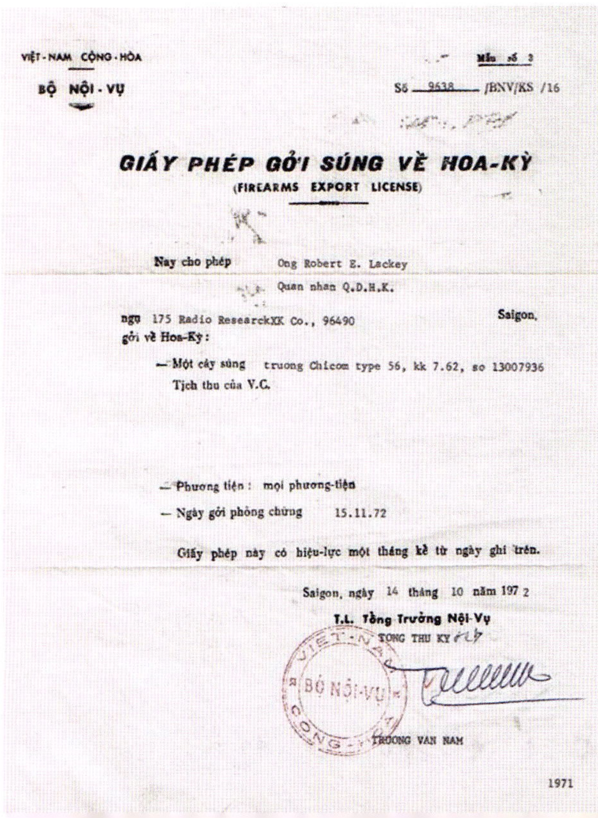

VIỆT-NAM CỘNG-HÒA

BỘ NỘI-VỤ

Mẫu số 3

Số 9638 /BNV/KS /16

GIẤY PHÉP GỞI SÚNG VỀ HOA-KỲ
(FIREARMS EXPORT LICENSE)

Nay cho phép Ong Robert E. Lackey
Quan nhan Q.D.H.K.

ngụ 175 Radio ResearckXX Co., 96490 Saigon,
gởi về Hoa-Kỳ:

— Một cây súng truong Chicom type 56, kk 7.62, so 13007936
Tịch thu của V.C.

— Phương tiện: mọi phương-tiện

— Ngày gởi phỏng chừng 15.11.72

Giấy phép này có hiệu-lực một tháng kể từ ngày ghi trên.

Saigon, ngày 14 tháng 10 năm 1972

T.L. Tổng Trưởng Nội-Vụ
TONG THU KY

BỘ NỘI-VỤ

TRUONG VAN NAM

1971

This SKS came out of the Cu Chi area of Vietnam and was brought home by a Veteran.

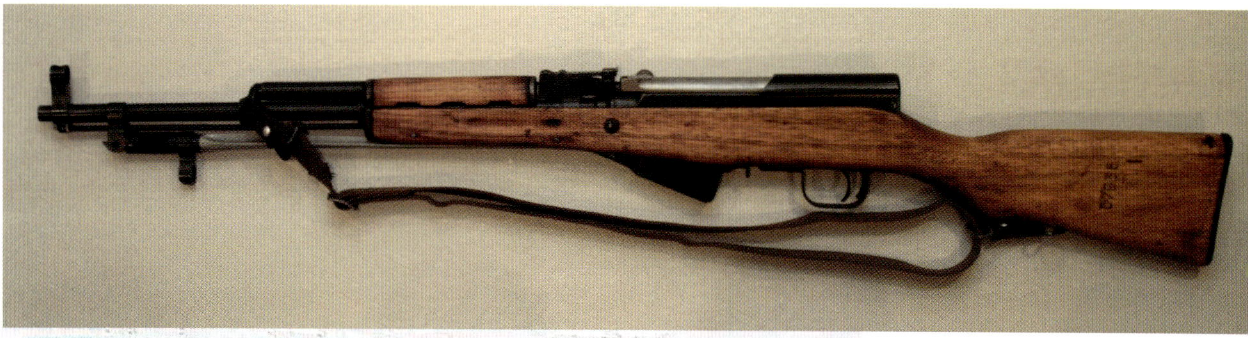

REGISTRATION OF WAR TROPHY FIREARMS

The original of this registration form will be retained by the person authorized possession. This registration is not transferable.

1. THEATER
REPUBLIC OF VIETNAM

2. NAME OF OWNER (Last name, first name, middle initial): Lackey, Robert, E.
3. SERVICE NO./SSAN: 510-46-2017
4. GRADE: E-6

5. ORGANIZATION
175th Radio Research Co., APO 96490

6. PERMANENT HOME ADDRESS (Street, City, State and ZIP Code)
711 S. E Wall Ave., Des Moines, Iowa 50315

Stamp: JOINT CUSTOMS GROUP Inspected and Passed SAIGON AREA DET

7. DESCRIPTION OF FIREARM

a. MAKE	b. TYPE (rifle, pistol, shotgun)	c. MODEL
CHICOM	CARBINE	TYPE 56
d. SERIAL NUMBER	**e. CALIBER**	**f. COUNTRY OF MANUFACTURE**
13007936	7.62mm	COMMUNIST CHINA

8. DATE: 18 May 71
9. TYPED NAME, GRADE, AND ORGANIZATION OF EXECUTING OFFICER: CECIL A. NEAL CPT, MPO PROVOST MARSHAL
10. STATION: BHAB
11. SIGNATURE OF EXECUTING OFFICER: Franklin R. Smith Jr.

Stamp: PROVOST MARSHAL OFFICE OFFICIAL 1st AIR CAVALRY DIVISION

DD FORM 1 OCT 51 603

U.S. GOVERNMENT PRINTING OFFICE : 1961 O - 971256

WAR TROPHY REGISTRATION/AUTHORIZATION

INSTRUCTIONS: This form will be prepared for all types of war trophies. Original to owner, 1 copy with trophy, 1 copy retained by issuing command. If war trophy is firearm, DD Form 603 will also be prepared.

Theater and inclusive period of service:
Vietnam: 21 Oct 70 - 22 Oct 71

Name of Owner (Last, First, MI): LACKEY, Robert E.
SSAN: 510-46-2017
RANK: SSG

Organization: 175th Radio Research Co. APO 96490
Permanent Home Address (Zip Code): 711 S.E. Wall Ave. Des Moines, Iowa 50315

DESCRIPTION OF WAR TROPHY

Describe item (If firearm indicate make, model, and caliber):
CHICOM CARBINE TYPE 56 7.62mm

Serial number or identifying mark (If firearm include country of manufacture):
#13007936 Communist China w/Bayonet

Date: 18 May 71
Type Name, Grade, Org of Authorizing Officers: RAFAEL G. MARTINEZ, SFC, HHC, 3rd Bde (Sep) APO 96490
Station: Bien Hoa Army Base

DD Form 603-1 - T.RA.

The capture paper is from my Uncle Russell Smith. He served on LST 970 in the Pacific during WWII. He was involved in landings on Chichi Jima and Okinawa. These pictures are from the Navy archives and show the ship unloading at Leyte and then on Ie Shima beach on Okinawa on 02/02/1946. The capture paper is dated April, 1946 and I remember my Uncle saying he got the rifle and bayonet on Okinawa so that makes sense. It is unknown where the rifle and bayonet are now.

Contributed by Stan Miller

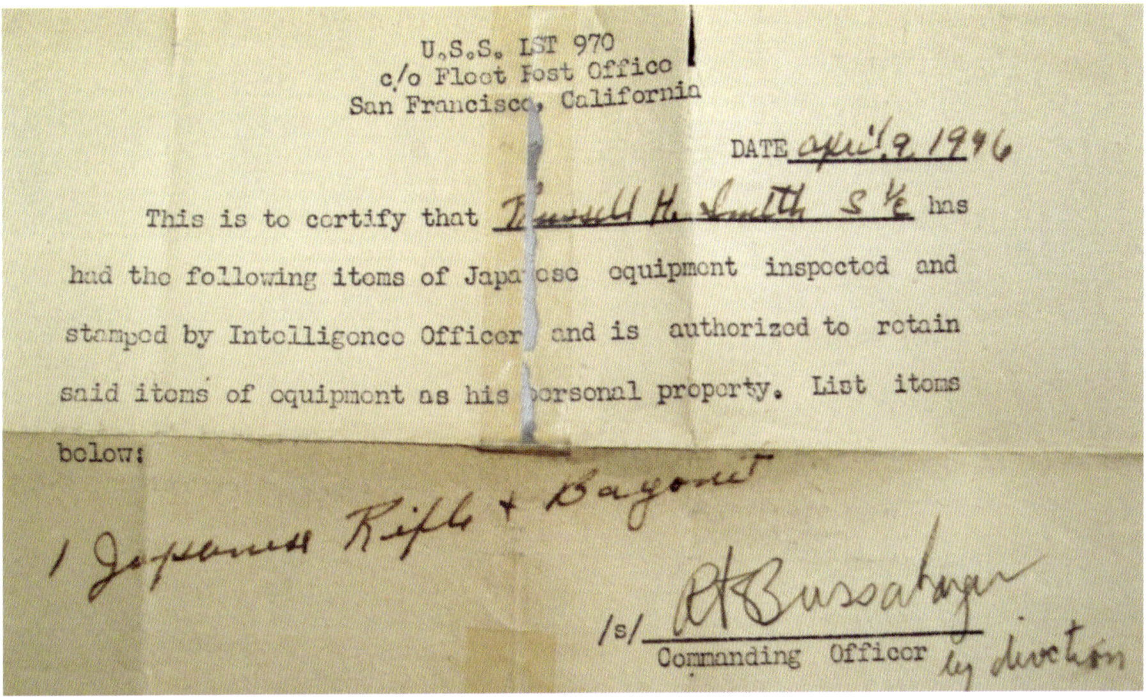

U.S.S. LST 970
c/o Fleet Post Office
San Francisco, California

DATE April 9, 1946

This is to certify that Russell H. Smith S 1/c has had the following items of Japanese equipment inspected and stamped by Intelligence Officer and is authorized to retain said items of equipment as his personal property. List items below:

1 Japanese Rifle + Bayonet

/s/ [illegible]
Commanding Officer by direction

Not a lot of story on this one, but interesting enough to show

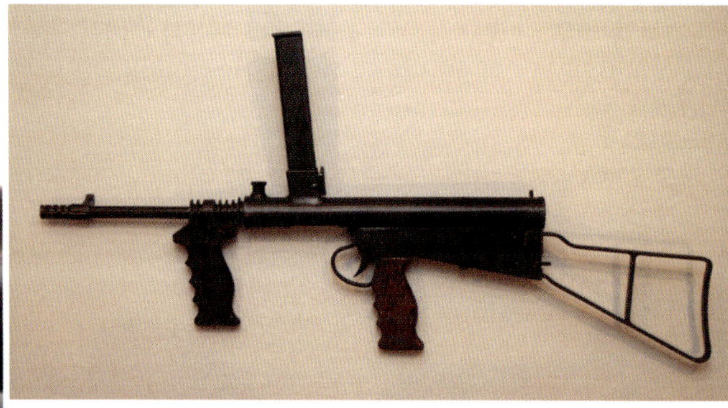

This Owens World War II weapon was given to me by Woody Johnson on March 22, 1957 at his house in Roseburg Oregon as a gift. He was a merchant seaman during the war and came by this weapon as a war souvenir of that conflict.

He gave it to me because he knew I had other hunting weapons and thought I might like to have it. He also worked for me part time at my tire store. He was a heavy drinker and got into lots of fights. He had a very short temper and this caused him a great deal of trouble. He was to put in a weeks work for me and he didn't show up for several days. When he came in to work he informed me he had been in the Douglas County Jail for the previous 4 days for disturbing the peace. In our conversation about this event I could tell he was still upset with the police department and he said lets go down to my house, I have something you might like to have.

He showed the weapon it to me and then he said he had better get rid of it as he has had urges to load it up and go down and clean out the Roseburg Police Department. So I accepted the weapon that day. He never did do anything that crazy as far as I know. I moved away shortly after receiving the weapon and I heard later from a friend Orley Milligan that he did stay out of trouble. Thank the Lord for that.

Arthur G. Van Alstine

Arthur G. Van Alstine

Manual for the Owen

The veteran's name is Rufus Arnold, he was from Washington, Indiana. I purchased the gun from his son Ron Arnold. I googled Rufus and he showed up in two articles, the first was about being on a winning high school basketball team. Then the second article I found was in the early 90's when several of the same team members got together. Mr. Arnold was in the grocery business but his son never made it clear if he owned a small local store or managed one. He said his father never spoke of the war or the gun but that he remembered the gun staying in a box in the hall closet from as early as he could remember.

When his father passed away he put everything in his attic. He only recently sold his house and was moving and decided to sell the gun. He didn't sell it directly, he used a broker and the broker put me in touch with Ron. One of the many things Ron sent me was the morning report, it mentions his father in three places. First it credits him with reducing casualties during one attack, it then mentions him for a purple heart. And lastly it mentions him for being in C battery which in the morning report seems to have seen the greatest amount of action.

Reading through the entire 339th morning log I have to guess it was taken between May 9th and June 4th 1945. In the log it states they were collecting troops, guns, equipment and munitions. But even that is a guess since the veteran passed away in 1999 and his son really knew very little about the gun but was willing to share with me all the information he had.
However I find the full day to day story of the 339th to be very interesting.
If you note the name on the paperwork you will find T/5 Rufus Arnold listed 3 times in "The Morning Report."

Morning Report of the unit

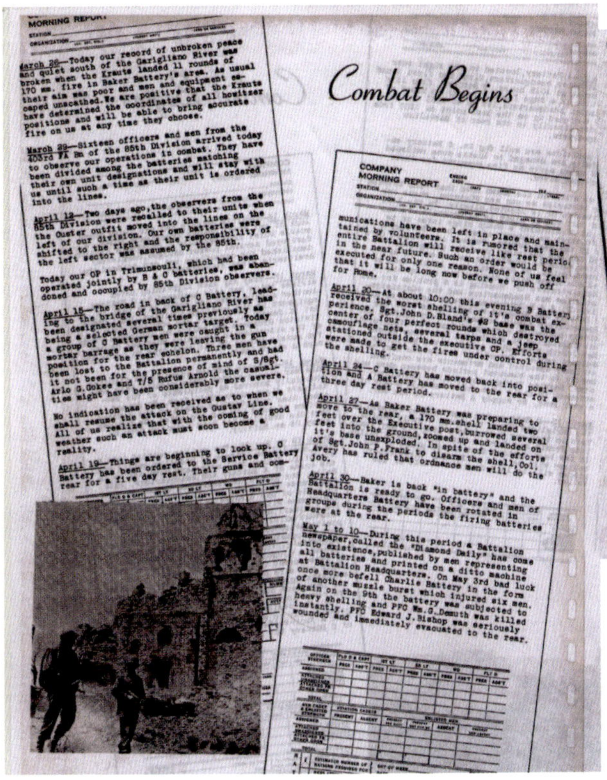

Today our OP in ... batteries, ... operated jointly by B & C batteries, ... doned and occupied by 85th Division observers.

April 15—The road in back of C Battery, leading to the bridge of the Garigliano River has been designated several times previously as being a selected German mortar target. Today a group of C Battery men were caught in a mortar barrage as they were leaving the gun position for the rear echelon. Three men have been lost to the Battalion permanently and had it not been for the presence of mind of S/Sgt. Arlo G.Gokee and T/5 Rufus Arnold the casualties might have been considerably more severe.

No indication has been received as to when we shall resume the attack on the Gustav Line. All of us realize that with the coming of good weather such an attack must soon become a reality.

April 19—Things are beginning to look up. C Battery has been ordered to the Service Battery rear for a five day rest. Their guns and com-

Purple Heart listed

BRONZE STAR MEDAL

Cpl. Edwin K. Griffith
Cpl. Joseph Gromacki
T/Sgt. Nicholas A. Grosso
Pfc. Walter W. Groth, Jr.
Cpl. Michael A. Guize
Cpl. Merlin J. Gutshall
Major Palmer Hanson
Tec. 5 Bailey O. Harvey
Tec. 5 Paul J. Hastings
Tec. 5 James H. Havens
S/Sgt. Edward F. Heuston
Tec. 4 John J. Hoffman
Tec. 4 John R. Innes, Jr.
1st Lt. Wayne A. Jaeke
S/Sgt. Charles W. Johnston
1st Lt. James E. Kane
Cpl. Harry B. Krause
Pfc. Sam Lamonica
1st Lt. Charles W. Leinker
Capt. Dale D. Levis
S/Sgt. Henry Liebowitz
Tec. 5 John C. Linder
1st Sgt. Carl C. Loftin
1st Lt. Elton M. Lovett
Capt. Frank D. Lowery
Tec. 5 Daniel R. Masciantonio
Tec. 5 Willard H. Masterson
S/Sgt. Joseph H. McCabe
Pfc. Murph McCaffity
Capt. Jewell L. Merritt
Tec. 5 Paul J. Miller

Cpl. Frank Mirabile
Tec. 5 Fred Misson
Tec. 4 Edward H. Mock
1st Lt. Edwin L. Molzahn
S/Sgt. Robert J. Muir, Jr.
M/Sgt. John H. Mulligan
S/Sgt. Milton G. Munson
Capt. Charles A. Nye, III
1st Sgt. Eliot W. Olson
Capt. Allan J. Oppenheim
Tec. 5 Albert E. Otterstrom
Cpl. Alvin R. Pappert
Tec. 5 Howard R. Phillips
Pfc. Stanley Pietroski
Pfc. Johnnie C. Potter
Pfc. Frank L. Pulice
Tec. 3 Paul E. Reece
1st Sgt. Floyd R. Reed
S/Sgt. Francis H. Reiley
1st Lt. Elwin T. Ricketts
Tec. 3 Stanley A. Roginski
Tec. 4 Chalmer A. Roles
Cpl. William S. Romanaski
Tec. 4 Stuart B. Rounds
Cpl. Troy G. Rowe
Tec. 5 John W. Rudesyle
Cpl. John S. Ruggiero
S/Sgt. John G. Safford
1st Lt. Howard Sails
Cpl. Michael Sakara
Sgt. Joseph V. Schember

Tec. 4 William H. Siegrist
Cpl. Sidney Simon
Major Harry B. Simpson
Tec. 5 Henry J. Sienkiewicz
Capt. Edward N. Sloane
Cpl. John J. Snee
Pfc. Marvin L. Snodgrass
Tec. 4 Jack A. Snow
Cpl. Doyle M. Snyder
Tec. 5 Jay O. Stradley
Sgt. Morris Strumer
S/Sgt. John M. Sullivan
Cpl. Edgar A. Terrell
Tec. 5 Edmond Terrell
Sgt. Sebastian J. Tiso*
1st Sgt. John J. Tooles
Tec. 4 Rupert C. Tucker
Tec. 4 Mark L. Verdier
Tec. 5 Vernie O. Walk
S/Sgt. Melvin E. Wall
S/Sgt. Daniel H. Wallace
Cpl. Thomas D. Walsh*
S/Sgt. Michael Wargo
2nd Lt. Wallace E. Waters
Tec. 5 Alfred D. Wattner
Sgt. Morton J. Weiss
Sgt. Julio Williams
Tec. 4 Lynford M. Wright
Cpl. Lowell A. Yarian

AIR MEDAL

2nd Lt. Odell Braswell
1st Lt. Henry Bruns*****
1st Lt. Walter E. Eschen
1st Lt. John T. Hayden********
1st Lt. Francis R. Moore*******
1st. Lt. Howard C. Wiggin******

ORDER of the PURPLE HEART

Tec. 5 Rufus Arnold
Tec. 5 Herbert P. Barger
Pfc. Shelia H. Bell
Tec. 5 Julius C. Bertoti
Pvt. Chester L. Bisbee
S/Sgt. John D. Bland
Sgt. Ciro L. Bologna
Pvt. Clarence A. Bowen
Pvt. Everett D. Bush

Cpl. Charles S. Callison
CWO Francis J. P. Clifford
Pvt. Gerard P Cloutier
Pfc. John A. Collins
Sgt. William T. Cunningham*
Tec. 4 Roy F. Currie
Pvt. Salvatore J. Divenuto
Sgt. Sidney Dobish
Pvt. Chester V. Druzbik

S/Sgt. Arlo G. Gokee
Pfc. Morris Goldrich
Tec. 5 Lewis G. Graham
Pfc. Earl R. Gurnee
Pvt. Charles S. Interrante
1st Lt. Wayne A. Jaeke
Pvt. Peter J. Jankowsky
Pfc. Manuel Johnson
Pfc. Walter J. Kendra

ORDER of the PURPLE HEART

Tec. 5 Rufus Arnold
Tec. 5 Herbert P. Barger
Pfc. Shelia H. Bell
Tec. 5 Julius C. Bertoti
Pvt. Chester L. Bisbee

Cpl. Charles S. Callison
CWO Francis J. P. Clifford
Pvt. Gerard P Cloutier
Pfc. John A. Collins
Sgt. William T. Cunningham*

His son tells me they were strafed by an ME109 and the truck he was behind rolled over on one of his legs and this was the reason for the purple heart.

A very nice 1937 S/42 Mauser

Picture of Rufus Arnold, probably at Fort Gruber in Oklahoma 1942.

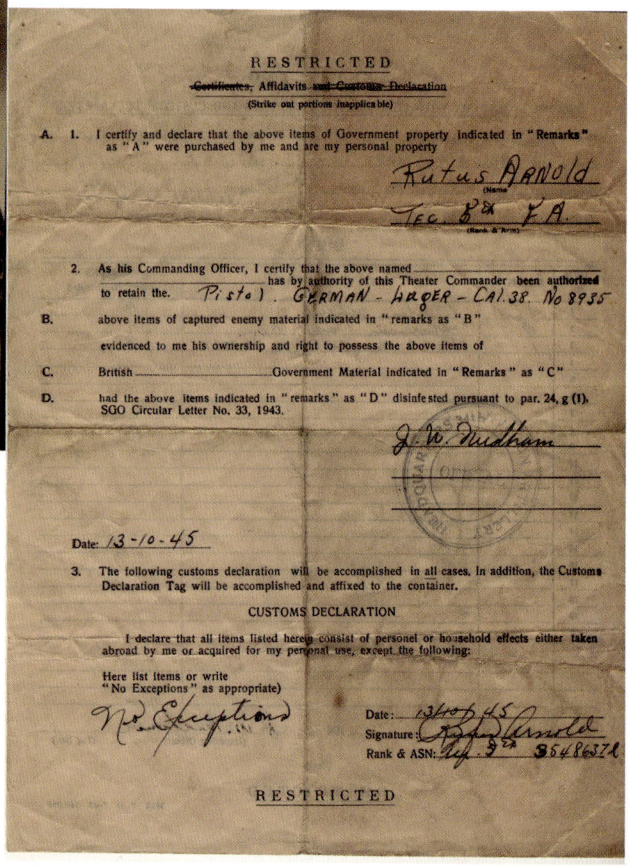

RESTRICTED

~~Certificates,~~ Affidavits ~~and Customs Declaration~~

(Strike out portions inapplicable)

A. 1. I certify and declare that the above items of Government property indicated in "Remarks" as "A" were purchased by me and are my personal property

Rufus Arnold
(Name)

Tec. 5th F.A.
(Rank & Arm)

2. As his Commanding Officer, I certify that the above named ______ has by authority of this Theater Commander been authorized to retain the. Pistol. GERMAN - LUGER - CAL. 38. No 8935

B. above items of captured enemy material indicated in "remarks as "B"

evidenced to me his ownership and right to possess the above items of

C. British ______ Government Material indicated in "Remarks" as "C"

D. had the above items indicated in "remarks" as "D" disinfested pursuant to par. 24, g (1), SGO Circular Letter No. 33, 1943.

J. W. Needham

Date: 13-10-45

3. The following customs declaration will be accomplished in all cases. In addition, the Customs Declaration Tag will be accomplished and affixed to the container.

CUSTOMS DECLARATION

I declare that all items listed herein consist of personel or household effects either taken abroad by me or acquired for my personal use, except the following:

Here list items or write
"No Exceptions" as appropriate)

No Exceptions

Date: 13/10/45
Signature: Rufus Arnold
Rank & ASN: Tec. 5th 35486372

RESTRICTED

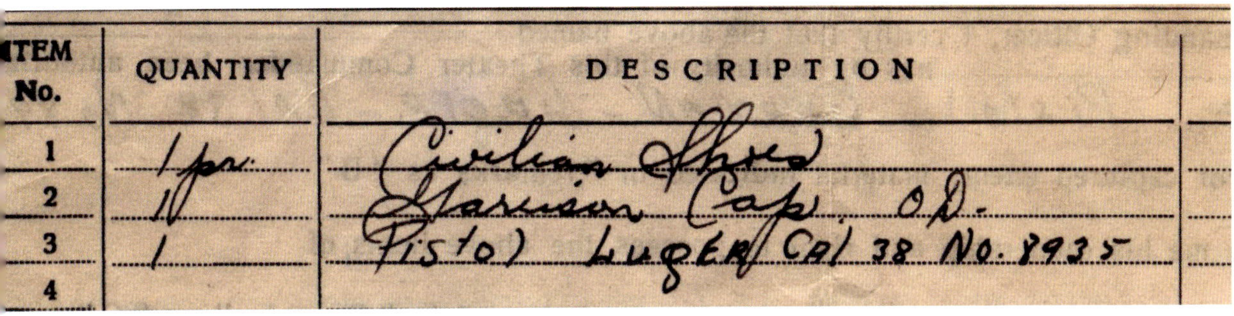

ITEM No.	QUANTITY	DESCRIPTION
1	1 pr.	Civilian Shoes
2	1	Garrison Cap, O.D.
3	1	Pistol Luger Cal 38 No. 8935
4		

An interesting thing on the bring back paperwork is that whoever did the paperwork got one digit wrong. I have talked to the son on the phone and have a letter from the son to verify it is the gun that was brought back but the customs document states it is 8935 but the gun is 6935 but looking at the magazine I can see how at a quick glance they may have gotten it wrong.

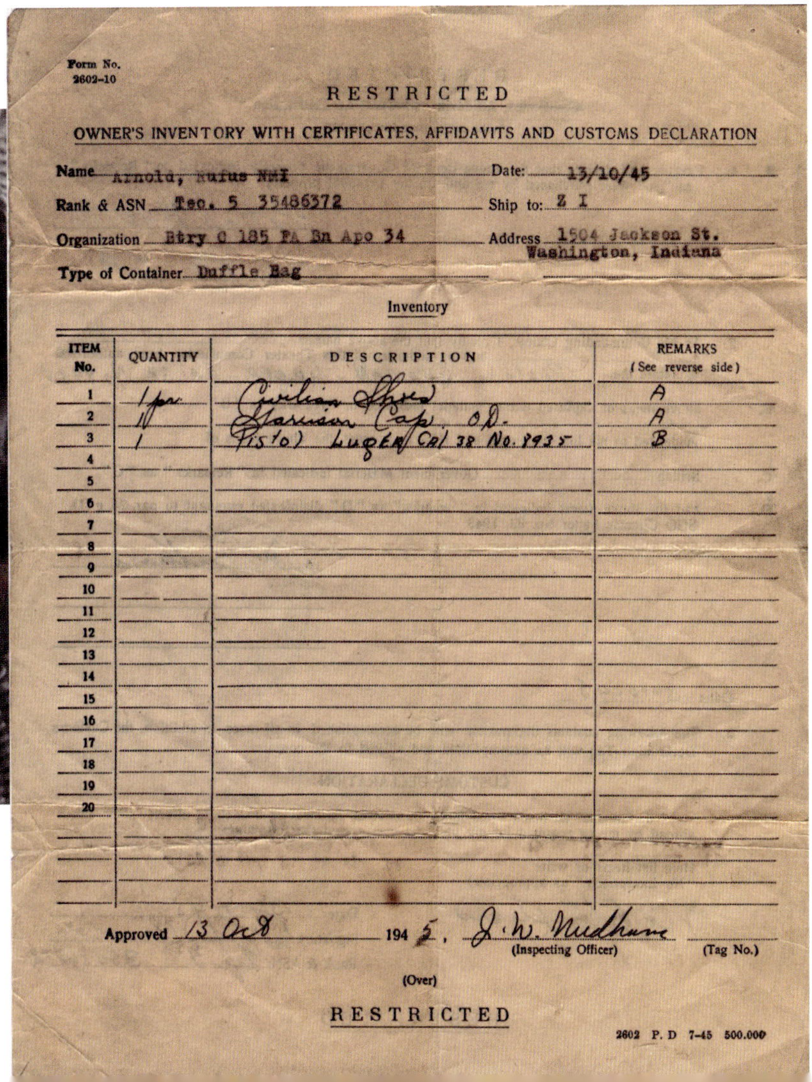

Form No. 2602-10

RESTRICTED

OWNER'S INVENTORY WITH CERTIFICATES, AFFIDAVITS AND CUSTOMS DECLARATION

Name Arnold, Rufus NMI — Date: 13/10/45

Rank & ASN Tec. 5 35486372 — Ship to: Z I

Organization Btry C 185 FA Bn Apo 34 — Address 1504 Jackson St. Washington, Indiana

Type of Container Duffle Bag

Inventory

ITEM No.	QUANTITY	DESCRIPTION	REMARKS (See reverse side)
1	1 pr	Civilian Shoes	A
2	1	Garrison Cap, O.D.	A
3	1	Pistol Luger Cal 38 No. 8935	B
4			
5			
6			
7			
8			
9			
10			
11			
12			
13			
14			
15			
16			
17			
18			
19			
20			

Approved 13 Oct 1945. J. W. Needham (Inspecting Officer) (Tag No.)

(Over)

RESTRICTED

2602 P. D 7-45 500.000

Edward Mattioli—Japanese Sword

This sword and a type 99 rifle was brought back by my late father.
The paper reads:
1182nd engineer refer maintenance detachment
APO 901 San Francisco Calif
Antonio Mattioli Tec/5 33840745
One (1) Japanese rifle 31 Cal and one (1) Japanese Saber (note: They spelled saber wrong on the paper)

My Dad was in the 40th Infantry Division. 24th Corps and he passed away in March 2006.

The scabbard has several bullet nicks in it.

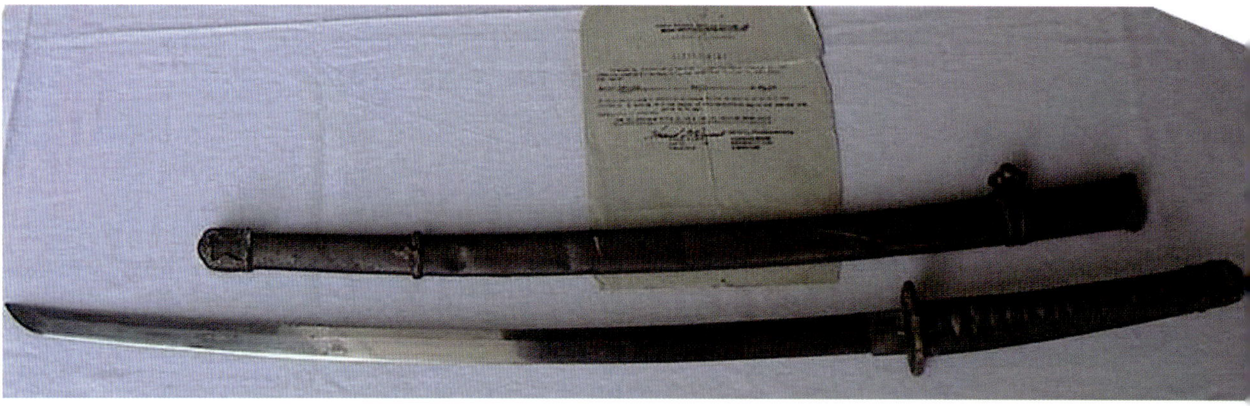

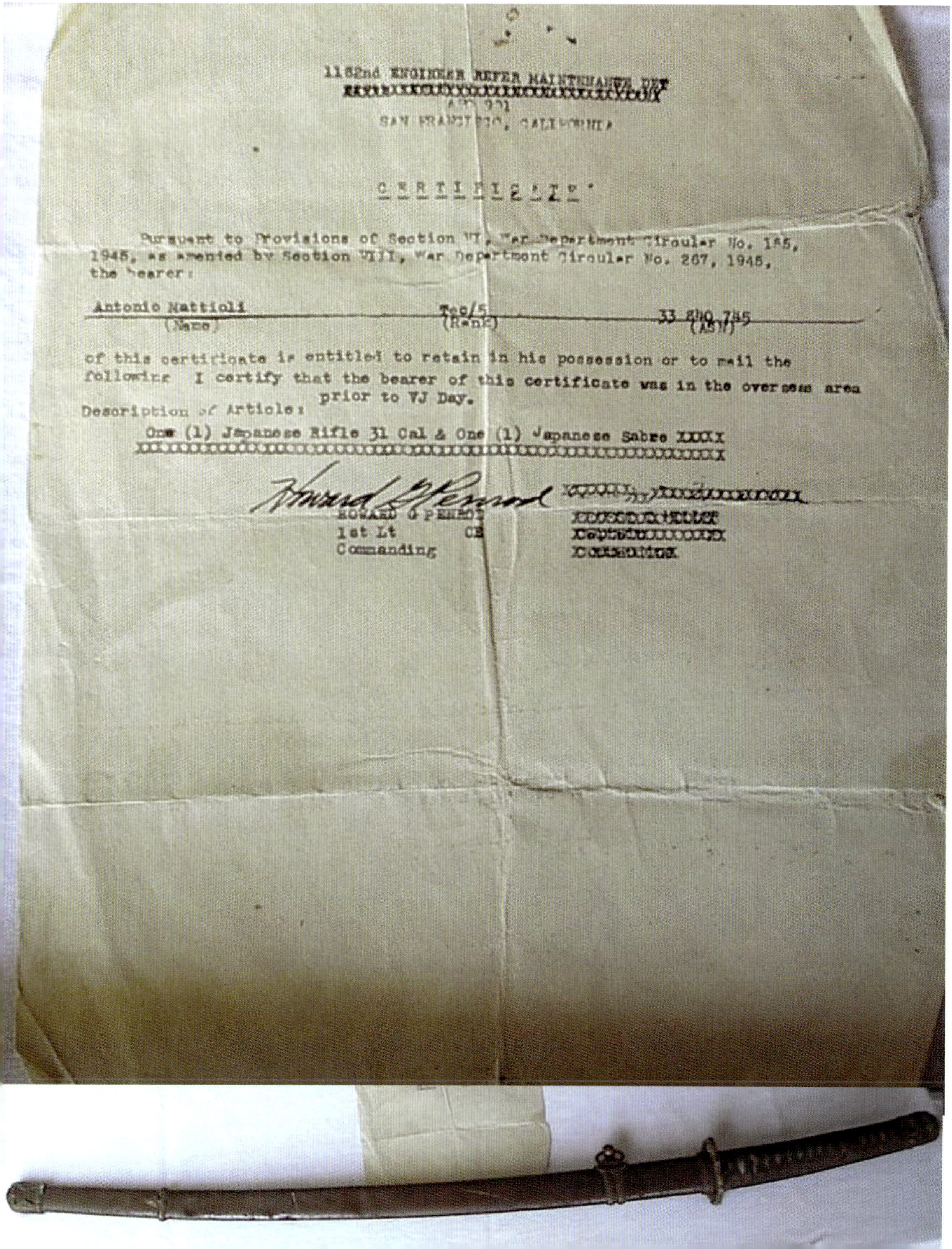

1162nd ENGINEER REFER MAINTENANCE DET
APO 921
SAN FRANCISCO, CALIFORNIA

CERTIFICATE

Pursuant to Provisions of Section VI, War Department Circular No. 155, 1945, as amended by Section VIII, War Department Circular No. 267, 1945, the bearer:

Antonio Mattioli (Name) Tec/5 (Rank) 33 840 745 (ASN)

of this certificate is entitled to retain in his possession or to mail the following I certify that the bearer of this certificate was in the overseas area prior to VJ Day.

Description of Article:

One (1) Japanese Rifle 31 Cal & One (1) Japanese Sabre

HOWARD G PENROD
1st Lt CE
Commanding

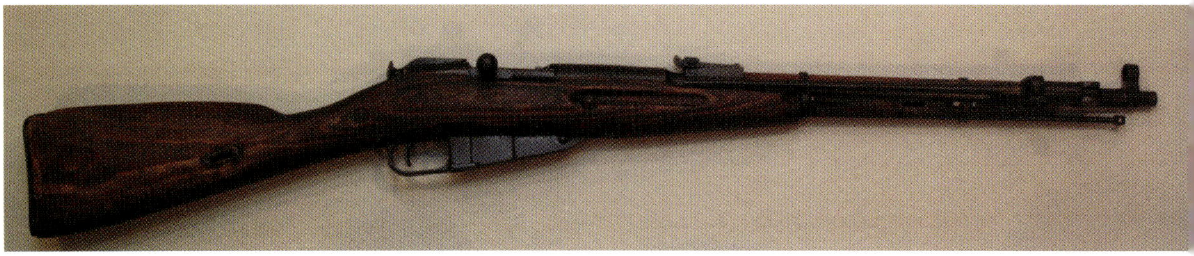

AFFIDAVIT
David John Wright
April 18, 1990

AFFIDAVIT

State of Oregon
County of Linn

I, David John Wright, BEING DULY SWORN, DEPOSE AND SAY:

1. WHEREAS, I, at the end of my service in the U. S. Navy River Patrol Force in the Mekong Delta area of the Republic of Vietnam, brought home with me, on July 21, 1967, a registered war trophy, to wit:

> "a Soviet Mosen-Nagent 7.62 mm carbine manufactured in 1944, serial number MP1427, with a Chinese Communist bolt, serial number J6263, which was manufactured in 1952, as confirmed from U. S. Army intelligence manuals in 1967", and;

2. WHEREAS, subject weapon was given to me by a U. S. Navy Junk Force advisor who recovered it during an ambush of Viet Cong guerillas in the U Minh Forest, approximately 30 miles south of Rach Ghia, in March 1967, and in view of its manufacturing history, it appears that the described weapon might well have been used in World War II by Soviets, in the Korean War by Chinese, and in the Vietnam War by Vietnamese Communists.

Further the affiant sayeth not.

David John Wright

Before me, a Notary Public, on this day personally appeared David John Wright, whose name is ascribed to the foregoing instrument, who swore to and subscribed his name thereto for the purposes and consideration therein expressed.

Given under my hand and seal of office this 18th day of April, 1990.

Sanna Krogh
Commission Expires on
10-09-93

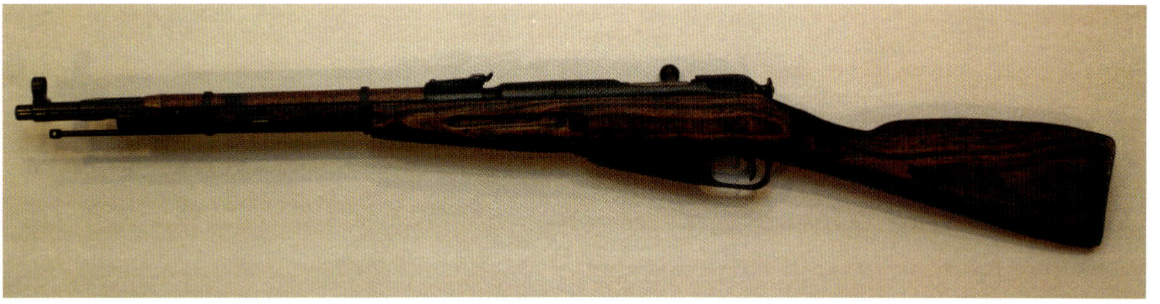

REGISTRATION OF WAR TROPHY FIREARMS

The original of this registration form will be retained by the person authorized possession. This registration is not transferable.

1. THEATER
Southeast Asia (RVN)/2 AUG 66 - 24 JUL 67

2. NAME OF OWNER (Last name, first name, middle initial)
WRIGHT, David J.

3. SERVICE NUMBER
58030[illegible]

4. GRADE
LCDR (O-4)

5. ORGANIZATION
United States Navy

6. PERMANENT HOME ADDRESS (Street, city, state)
194 So. Lake St.
Albany, Oregon, U.S.A.

7. DESCRIPTION OF FIREARM

a. MAKE
Soviet [illegible]

b. TYPE (Rifle, pistol, shotgun)
Carbine

c. MODEL
1944

d. SERIAL NUMBER
Breech MP1427
Bolt J6263

e. CALIBER
7.62mm

f. COUNTRY OF MANUFACTURE
Russia

8. DATE
1 JUL 67

9. TYPED NAME, GRADE, AND ORGANIZATION OF EXECUTING OFFICER
B. A. YOUNG, LT, USN, COMRIVPATFOR (CTF 116)

10. STATION
RIVPATFOR, Can Tho, (RVN)

11. SIGNATURE OF EXECUTING OFFICER
B. A. Young

DD FORM 1 OCT 51 603

FPC-Japan

OFFICIAL PROVOST MARSHAL OFFICE USARPAC APO

RECEIVED

WAR TROPHY REGISTRATION/AUTHORIZATION

INSTRUCTIONS: This form will be prepared for all types of war trophies. Original to owner; 1 copy with trophy; 1 copy retained by issuing command. If war trophy is a firearm, DD Form 603 will also be prepared.

Theater and inclusive period of service in overseas command
Southeast Asia (RVN)/2 AUG 66 - 24 JUL 67

Name of Owner (Last, First, MI)
WRIGHT, David J.

Service Number
580300

Grade/Rank
LCDR (O-4)

Organization
United States Navy

Permanent Home Address
194 So. Lake Street
Albany, Oregon, U.S.A. 97321

DESCRIPTION OF WAR TROPHY

Describe Item (If firearm, indicate make, type - rifle/pistol)
Carbine, 7.62mm, Soviet MOSIN-NAGANT Model 1944

Serial Number or Identifying Mark (if firearm, include country of manufacture)
BREECH - MP1427/BOLT - J6263

Date
1 July 1967

Typed Name, Grade, Org of Authenticating Officer
LT B. A. YOUNG, USN, COMRIVPATFOR (CTF 116)

Station
RIVPATFOR, Can Tho, RVN

Signature
B. A. Young

DD Form 603-1, 1 Jul 65

The 6.5 mm Japanese bullet Ralph was shot with on April 15th, 1945. He passed this bullet into a bed pan during one of his visits to a VA hospital in 1952.

Staff Sgt. Ralph C. Bigley - US Army, Americal Division, South Pacific.
As Staff Sgt. Ralph C. Bigley slowly regained consciousness during the early morning light of April 15th, 1945, he thought he must be in Heaven. He was lying comfortably inside a tent instead of sleeping on the rough open ground among rocks, coconuts and rats or standing neck deep in water in a foxhole on watch. He wasn't dodging bullets or mortar fire either, just lying alongside other soldiers peacefully relaxing on a warm, tropical morning on a remote island in th Pacific. As the pain started to set in, though, he realized the other soldiers lying near him weren' moving or talking. They weren't even breathing. Then it hit him….he was alive, but was in the tent where they placed the dead soldiers until the burial detail could get to them! Ralph began tc yell for medics. He quickly found himself leaving Heaven and entering what could only be described as Hell for the next 37 years, thanks to that one small bullet a Japanese soldier shot int his body just a few hours earlier.

Ralph took his pre-induction physical at Ft. Leavenworth, KS, on 7 April, 1943, was inducted into the U.S. Army three days later, and arrived at Camp Robinson, AR, on 28 April, 1943. On the back of an Army postcard he mailed home he wrote "Dear Mom, We have to drill 16 hours a day. I am in the infantry. Those guys at Leavenworth sure do lie. They told me I would go in th Salvage Dept. Well, I guess I can take it. Love to all, Ralph." Ralph completed infantry boot camp and was assigned to Company E of the 2nd Battalion, 164th Infantry, Americal Division. The Americal was actually the 23rd Division, but was one of a handful of Army divisions referred to by name rather than number during WWII. He embarked for the South Pacific on 3 October, 1943, and arrived at Viti Levu, Fiji Islands on 25 October, 1943. There he joined the Americal as it was refitting and training new replacements following it's combat on Guadalcana

The Americal soon entered combat on Bougainville Island in the Solomons. During nearly a full year there they faced numerous enemy artillery and infantry attacks, many times within their own their perimeter, while at the same time sending out combat patrols to locate, harass, and kill the Japanese soldiers trying to drive them from the island. Life wasn't easy for Ralph or his fellow soldiers while fighting overseas. In addition to fighting an enemy that very much wanted to kill them, they battled extreme weather conditions, endured primitive living conditions, and were constantly reminded of the peril they were in by seeing buddies wounded or killed.

In January 1945 the Americal entered combat on Leyte Island in the Philippines. On 1 March, 1945, Ralph was involved with action that earned him the Bronze Star. While his unit was still preparing emplacements on a hill that it had occupied a few hours earlier, the Japanese opened fire with machine guns, wounding one of the soldiers. Ralph crawled forward under the cross-fire until he reached a point from which he could determine the enemy positions. He passed back information directing mortar fire, remaining in place without concealment and drawing shots from an enemy sniper until one enemy machine gun had been knocked out and the rest routed. He then returned to assist in evacuating the wounded man. Two days later Ralph was admitted to an aid station. His physical condition had deteriorated tremendously, probably due to combat fatigue, stress, and the living conditions in the Pacific.

On 9 April, 1945, Ralph's unit was sent to nearby Cebu Island and participated in the battle for Babag Ridge. During attacks over a series of grassy, open-topped hills the day after receiving a "Dear John" letter from his wife Ralph was shot and critically wounded. He had been on Cebu for less than one week. Ralph's sister Judy remembers her parents talking about Ralph receiving the letter one day and then getting wounded the next. They always thought the letter he received affected him mentally and he was not as alert as he could have been when shot. My mother Judy recently wrote "A Mother's love can be so strong! On the morning when he awoke in the tent with the dead soldiers Mom was dreaming about him. She later told us about it and could hear him calling out for help." Help came and Ralph began his journey home, though it wouldn't be quick enough to suit him.

Ralph would undergo two operations in as many months while in the Philippines before being shipped out to the United States when doctors determined he was not likely to survive unless he received better treatment. Ralph spent the next ten months at various hospitals in the USA while undergoing two more operations. Surgeons never retrieved the bullet because it was too close to Ralph's spine and they feared it could paralyze him if they tried to remove it. The decision was finally made to leave it in his body.

Ralph Bigley's bring-back story doesn't involve a captured battle flag, rifle, pistol or other typical spoils of war soldiers bring back from war. Ralph's bring-back item was a single, small chunk of metal. Specifically, he carried a 6.5 mm bullet back home in his body courtesy of a Japanese soldier desperately fighting for his life on 15 April, 1945 while trying to stem the American advance. It wasn't until 7 years later Ralph or anyone else was able to physically see the bullet and hold it in their hands. During one of his visits to the veteran's hospital in Fayetteville, AR, in 1952, Ralph passed the bullet through his urinary tract. The doctors and nurses were amazed that such a thing could happen! I'm sure Ralph was not only amazed, but relieved as well to finally have that chunk of metal out of his body. I have that bullet now and it is the centerpiece of my Japanese military collection. I will treasure that small, single bring-back item more than any other single piece of military equipment or gear I might be able to collect. It may be small, but in my mind it stands as a huge testament and reminder of the price my Uncle paid for his service to our country.

Submitted by Scott Burger

Ralph somewhere in the Philippines (Center back row)

My father was a young 17 year old Marine stationed in China in 1945. At war's end their job was to guard Japanese prisoners. They did that until 1946. He said inside the camp the Japanese ran things like a military base, surrounded by guards and fence. My Dad said that they were all supposed to be enlisted men but that a hand full of them were older and distinguished looking and were definitely in charge. The other Japanese treated them with the utmost respect. They were polite to the Marines and the Marine guards left them alone. When they left for repatriation, the Japanese "corporal" in charge gave my Dad a Mauser HSc, serial number 822639.
How did he get a German pistol? Where did he get it? How did he manage to keep it as a prisoner? My Dad said the Marine officers let him keep it to control his POWs. Who was the "corporal"?

When the Marines pulled out of China in 1946 my Dad was in the hospital with measles and jungle rot on his hands and feet. They came into the hospital and announced that anybody that could walk out should board the ship now. Anyone left would become wards of the Red Chinese that were taking over the area.

My Dad could walk and he came back. I wonder what happened to the ones left behind. When my Dad died, the jungle rot on his hands and feet was starting to finally get a little better. My Dad's name: Lewis G. Kester, Sgt., U.S.M.C. 1945-1949—I got the HSc when Dad died in 1985.

Inspection mark on right of end of barrel

A Browning Hi-Power with Vietnam certificate.

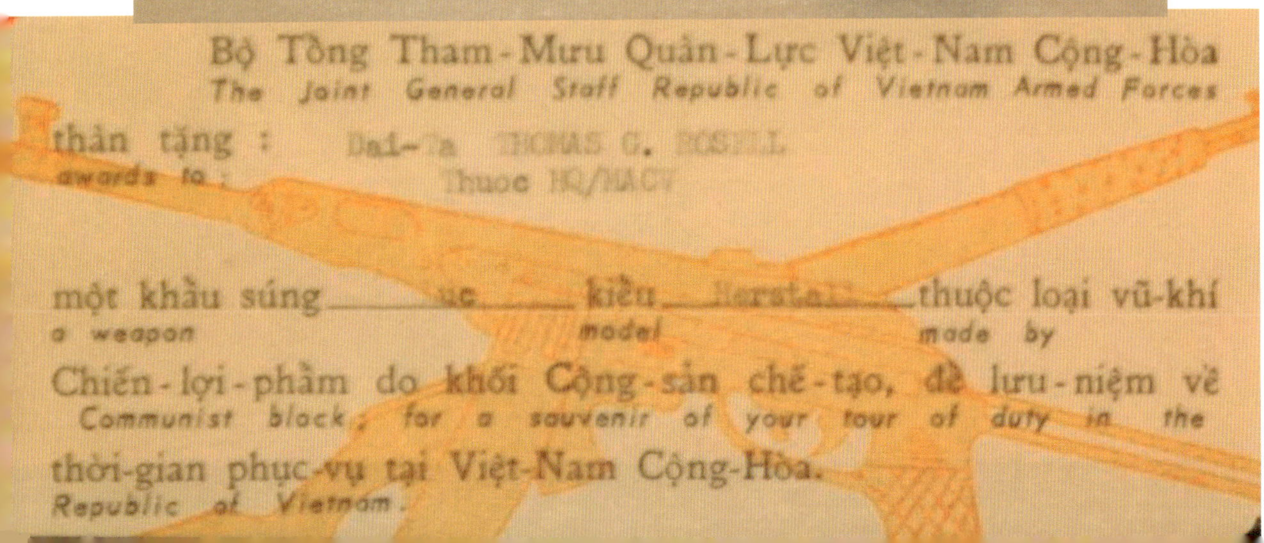

Oscar Piowaty—Cluster bomb

My Grandfather was a Captain in the Merchant Marines. He had a very colorful career starting in the 1920s on a side wheeler traveling up and down the Yangtze River.
He was torpedoed by the Germans and spent three days in a life raft and then later he was cluster bombed by the Japanese. One cluster bomb landed behind him on the partial open bridge. Luckily for him (and me) it was a dud. The crew dewatted it and presented it to him as a lucky gift. He gave me the bomb one day as he told me of his sea tales. Unfortunately he passed away when I was six. It was that bomb, his stories and my Fathers antique shop walk-ins that started me collecting Japanese militaria.

Contributed by Jareth Holub

I obtained this gun from the family of the man who take it directly from the German during WWII.

This police Luger has a WW1 dated 1915 holster that has been modified to police use. The G date is not common in being a police Luger, only 20,000 in the eagle 63 were made in that year; and of those 20k, only 600 were made for police use. A matching magazine is a plus, that is for sure, although found more in police guns than in army guns. It may have received the sear safety when produced or just a bit later, but very close to the same time frame.

Hershel Hubert "Poppy" Hoover —Nazi Dagger

My wife, Jamie Hoover-Miller's grandpa, Hershel Hubert "Poppy" Hoover was an infantryman in Company A, 71st Infantry Regiment of the 44th Infantry Division. He began his war landing in Cherbourg and fought all the way into Mannheim, Germany. On the 44th Website, they talk about a battle during the Bulge that occurred near Remling, France. The 71st Regiment repelled repeated attacks from the German SS regiment "Gotz Von Berlichegen" where the Germans screamed "Die Yankee Bastards" as they charged. The unit received a presidential unit citation for that. He brought back a 2nd Pattern Luftwaffe dagger and a Beretta pistol. The dagger is pictured here along with a photo of "Poppy." He said that while on guard duty at the hospital in Mannheim, a German Captain came up to him and surrendered. He said the officer said something like "I surrender this to you" and handed him the dagger as well as the pistol. My Father in law received the dagger and his brother got the pistol. Poppy is the one on the left of the photo, not the one in the M41 jacket. The photo is backwards. The patch is actually on the left shoulder of the other soldier.

Contributed by Stan Miller

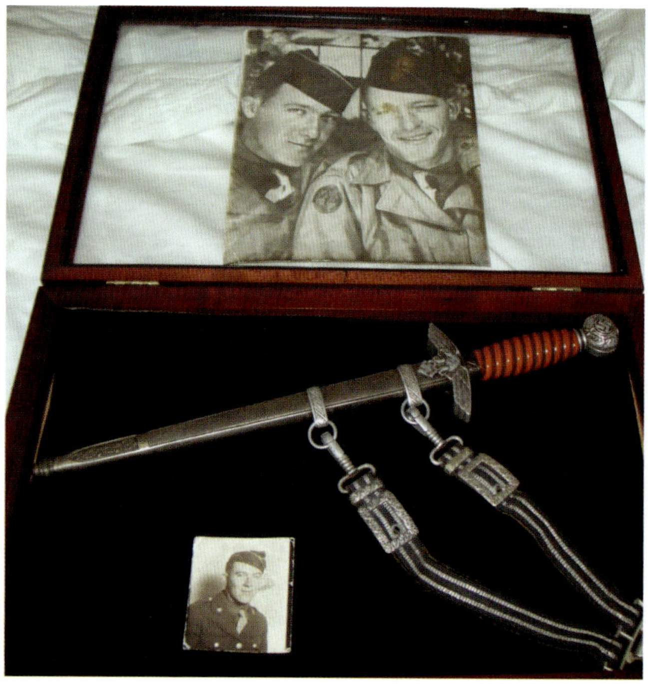

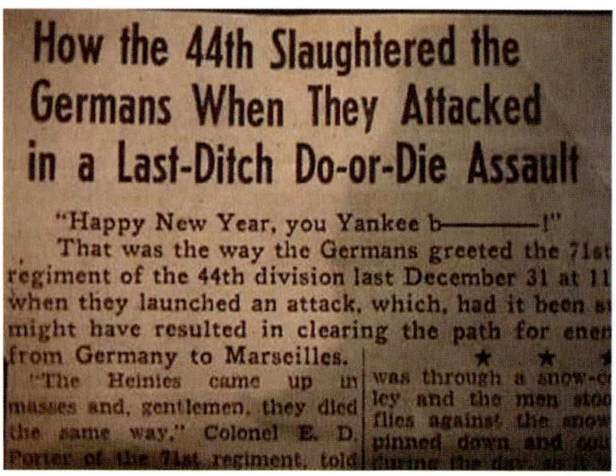

How the 44th Slaughtered the Germans When They Attacked in a Last-Ditch Do-or-Die Assault

"Happy New Year, you Yankee b——!"

That was the way the Germans greeted the 71st regiment of the 44th division last December 31 at 11 when they launched an attack, which, had it been s might have resulted in clearing the path for ener from Germany to Marseilles.

"The Heinies came up in masses and, gentlemen, they died the same way," Colonel E. D. Porter of the 71st regiment, told

★ ★ ★

was through a snow-c ley and the men stoo flies against the snow pinned down and cou during the day, ...

Here is a picture of Hershel "Poppy" Hoover holding the Luftwaffe dagger he picked up in WWII. He will be 87 this month (August 2010).

I own a Japanese Browning 1910 pistol rig brought back by a sailor on an attack transport ship. It's accompanied by a capture document that references the island of Peleliu. I purchased the rig from the Veteran's son and wife here in Florida. There are several other items listed on the invoice and the last time I spoke with the son (several years ago) they were in Long Island, N.Y. in the possession of the Veteran's daughter. This is also some capture paperwork with the pistol.

I certify that prior to my departure fromPeleliu Island, insufficient time was available for proper clearance of items listed through local intelligence officer.

Sidney Sadovnick
SIDNEY SADOVNICK
810-70-39, MoMM1/c

TROOP COMMANDER
U.S.S. CLERMONT (APA 143)

Date *Nov. 21, 1945*

SADOVNICK	*SIDNEY*	—	*810 70 39*	*MoMM 1/c*	*U.S.N.R.*
Surname	First	Initial	Serial No.	Rate	Service

the bearer of this certificate, is entitled to retain as his property the following:

ITEM	DESCRIPTION OF MATERIAL
	BAYONET (79615)
(1)	*RIFLE 25 CALIBRE - JAPANESE (196555 4)*
(1)	*P[illegible]OL 32 CALIBRE - JAPANESE (422962)*
(1)	*KNIFE - JAPANESE*

By direction of
The Commanding Officer.

(Signed) *J. C. Mericantante*
J. C. MERICANTANTE
2dLt, USMC.
Rank and Service

(It is permissible to mail the above described material to the United States PROVIDED it does not consist of firearms capable of being concealed on the person, or articles not permitted to be retained as souvenirs under existing directives.)

Edward Mattioli—Japanese T99 in shipping container

Complete in its original field-made shipping container is a Nagoya type 99 that was mailed home as a souvenir. This particular item was mailed from APO 468 (Leyte, PI 1944) The rifle inside the box is a 5th series rifle that has an "X" on the chrysanthemum rather than grinding, denoting surrender in the field. The rifle is as found 60+ years later inside the box. Typical of the time period, the bolt is from a later style type 99 rifle.
Thousands of boxed rifles like this came home from the Pacific Theater in field-made containers such as these. The complete shipping containers are exceptionally rare finds today and are of special interest in and among themselves.

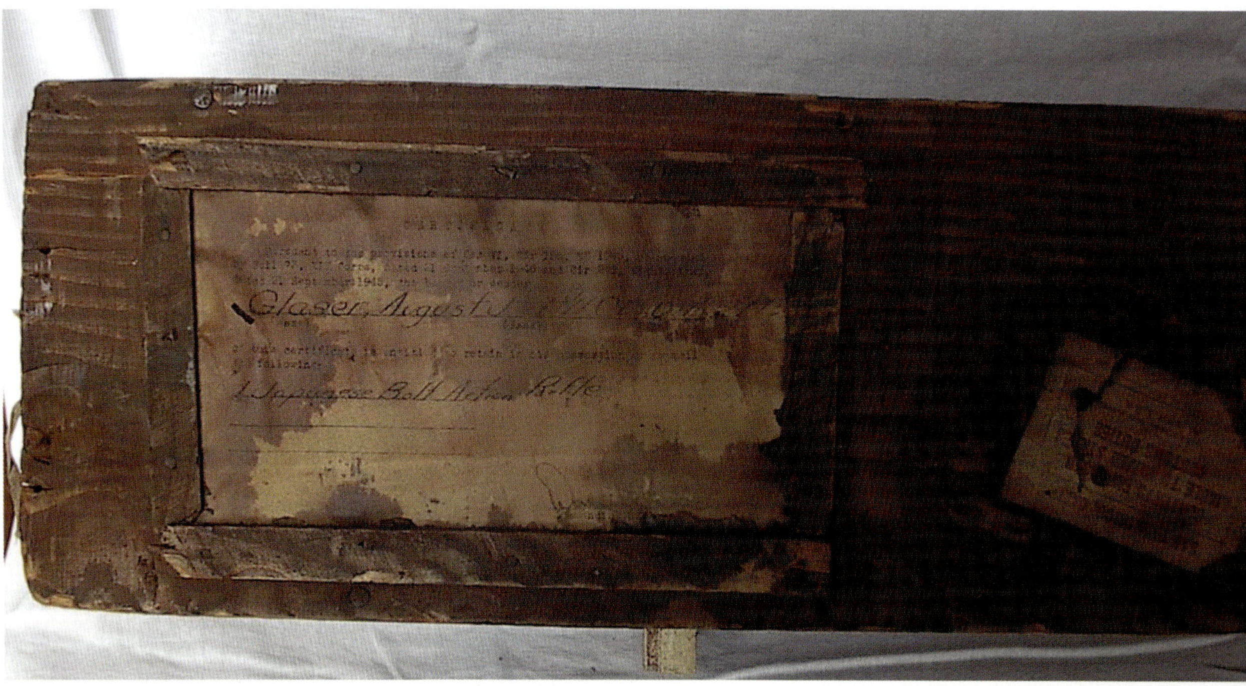

Complete in its original field-made shipping container, the label states:

LT. A J Glaser
46th Engr Cons. Bn. Co. C
APO 468 c/o PMSF, Calif

To: Helen Jackson
W Arrow Hwy
Upland, Calif

Box contents 1 Japanese bolt Action Rifle

Two elderly gunsmiths my dad knew a few years back from our antique store were at my mom's booth at an antique mall one day and they struck up a conversation. Difference was that the older one had hearing aids, and the younger one was nearly blind. When one would strike a conversation, the other would answer something completely random...unknown how they got along. Now, they had the revolutionary technology of Flexon and hearing aids! But, aside from the point, we told the older one about some antique cap pistols and a relic CG Bonehill shotgun we had, and he was interested. We gave him a card and my mom also mentioned I was a WWII militaria collector and interested in Lugers and Japanese swords (along with I'm gonna be a "man of the sea"). With this, he told us a story:

He had enlisted in the Coast Guard in 1945 because of a condition that would've rejected him from many other branches. Right after he graduated, the war ended and he was sent for occupational duties in Japan. While he was overseas, he traveled around there a bit, saw what postwar Japan had to offer, and like the souvenir hunter of the stereotypical GI, wanted something. What he wanted was what his friends and fellow members brought back: A Japanese sword. However he didn't know of the warehouses stored with swords and most of his shipmates had sent theirs home.

While talking to a fellow shipmate one day, however, he mentioned he wanted a sword and wondered where to get one. This is when his friend told him he had brought one onboard. He didn't really want it, he just picked it along the way! But he also didn't want to give it away, so he went about the decks bartering for the currency of the day, cigarettes! He finally was able to secure four packs of Lucky Camels and his friend gave him the sword for the packs.

This is basically where the story ends. He told me that the scabbard's lacquer was chipping, so he repainted that and there was writing on the blade. Nakago or blade, unknown. He said he'd give us a call about the guns and the sword later to possibly sell or trade...but that was two months ago. I'm tempted to call him, but I don't want to bother him.

Many folks have heard of the locally made guns in Vietnam, they are not rare, but not a common find, especially with paperwork.

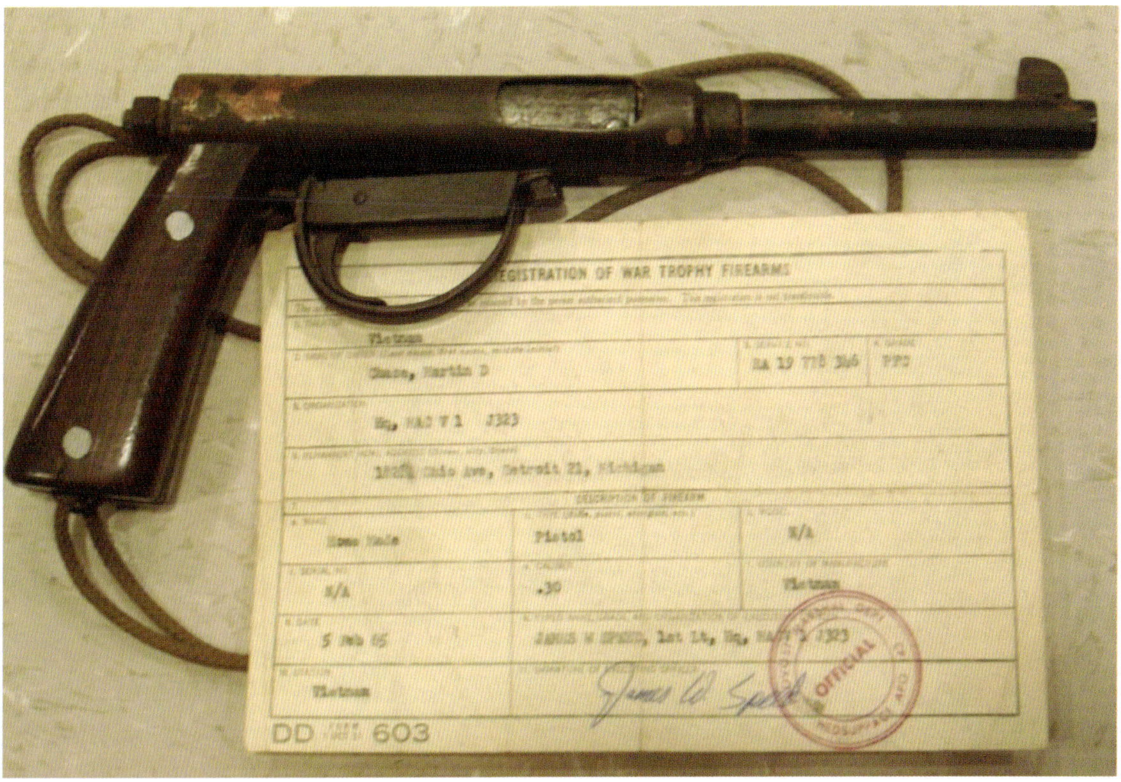

Two Vietnam bring backs, the zip gun above and another smoothbore rifle or shotgun

Joshua Beckett—Desert Storm Iraqi items

Here are Iraqi items from Desert Storm, these are the only ones that I have stories with, and they were picked up by an MP. The khaki uniform came from the highway of death, along with the map-case which was picked up in a crate full of map cases in a transport truck on the same highway. The uniform is made by Hyundai, but the helmet came from the Kuwaiti side of Khafji. The beret was also picked up in Khafji and originally had a badge on the front but it was removed/stolen by a souvenir hungry GI at the airport from the guy who originally brought it back. The MP who picked it up told me he took it because it appeared to be of better quality than the other berets he saw, its possibly a captured/imported beret. The mustard color berets were used by supply troops.

This Tokarev pistol was presented to General John H. Hay, former commander of the First Infantry Division in Vietnam from February 1967 to March 1968. Also pictured is Gen Hay's personal Signal Operating Instructions (SOI), or also called a call sign book.

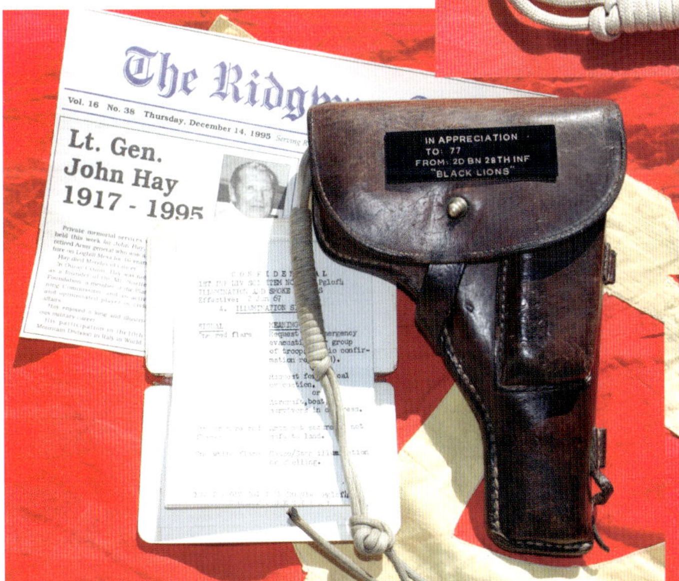

My grandfather brought back a ground mum, bolt mismatch, 1st variation T44 Arisaka which I have in my possession. He was an Army staff sergeant and was a BAR rifleman. He fought in New Guinea and the Philippines, I believe he was in A company, 20th Infantry Regiment, 6th infantry division. Unfortunately he died when I was 2 so I don't know the whole story behind how he got the bring back. He apparently brought back a sword too, but my grandmother sold it because it scared her.

Notice that the mum is ground, common to find, and is believed that it signifies that it was to keep honor for the Emperor by removing the crest / mum.

Notice the muzzle cover, as several experts on the Japanese forum thought it looked odd. However, it is just as my grandfather brought it back like this.

See the dust cover, many were lost, thrown away, it is assumed that the soldiers were told not to use them, but the Japanese collecting fraternity does not believe this theory.

A FEW WORDS ABOUT MY LUGER AND HOW I OBTAINED IT.

IN JULY 1942 I ENLISTED IN THE U. S. ARMY. AT THAT TIME THE BASE PAY WAS $21.00 PER MONTH AND AFTER INSURANCE AND LAUNDRY, THE NET AMOUNT WAS ABOUT $12.00. I ASKED IF THERE WAS ANYTHING THAT PAID EXTRA AND WAS TOLD THAT PARATROOPERS WERE PAID AN ADDITIONAL $50.00 AFTER THEY QUALIFIED. I DID NOT KNOW WHAT A PARATROOPER WAS BUT I SAID I WOULD JOIN.

I WAS SENT TO TOCCOA, GA. , A BRAND NEW CAMP WHERE THEY WERE FORMING A NEW PARACHUTE REGIMENT, TO BE KNOWN AS THE 506 PARACHUTE INFINTRY REGIMENT. AFTER A LOT OF TEST, I WAS ASSIGNED TO COMPANY "F", 2ND BATTALION.

I WAS ISSUED A NEW M-1 RIFLE AND PARATROOPER BOOTS. WE DID PARACHUTE AND INFINTRY TRAINING UNTIL DEC 1, 1942 AND THEN WALKED OUT THE FRONT GATE OF CAMP TOCCOA AND WALKED TO ATLANTA, GA., A DISTANCE OF 120 MILES IN 3 DAYS. WE THEN TOOK A TRAIN TO FORT BENNING, GA, WHERE WE DID MORE JUMP TRAINING AND 5 JUMPS TO GET OUR PARACHUTE WINGS.

AFTER MORE TRAINING AND JUMPS, WE WERE SHIPPED OUT OF NEW YORK TO ENGLAND IN SEPT., 1943 AND CONTINUED TO TRAIN IN ENGLAND THRU JUNE 5TH, 1944.

ON THE NIGHT OF JUNE 5TH, 1944 WE WERE LOADED ON C-47 TROOP CARRIER PLANES AND TOOK OFF TO GATHER IN FORMATION WITH HUNDREDS OF OTHER C-47 PLANES TO SPEARHEAD THE INVASION OF FRANCE. D-DAY WAS JUST ABOUT TO HAPPEN.

MY COMPANY "F" WAS TO JUMP ON AND CAPTURE A SMALL TOWN ON THE NORMANDY COAST. WE WERE TO KILL OR CAPTURE THE GERMAN SOLDIERS IN THE TOWN AND PLACE LAND MINES ON A ROAD THAT RAN FROM THE TOWN, DOWN TO THE BEACH. THIS WAS TO KEEP TANKS AND OTHER VEHICLES FROM BEING ABLE TO GET TO THE BEACH.

OUR ASSIGNED MISSION DID NOT WORK. OUR PILOTS HAD NEVER FLOWN COMBAT SO WHEN WE STARTED GETTING GROUND AND ANTI-AIRCRAFT FIRE, THE PILOTS BROKE FORMATION AND THE MEN IN MY PLANE LANDED VERY NEAR THE CHURCH IN THE CENTER OF ST MERE EGLISE. AS I CAME OUT THE DOOR OF THE PLANE, I COULD SEE THE CHURCH JUST IN FRONT OF ME AND COULD HEAR THE CHURCH BELLS RINGING. ALSO TWO BUILDINGS WERE BURNING WHICH MADE MOST OF THE TOWN VERY BRIGHT. ALSO, TRACER BULLETS WERE COMING UP SO THICK THAT THEY LOOKED LIKE STEPS THAT YOU COULD WALK DOWN ON. WE KNEW THAT WE WERE ONLY SEEING ONE OF EVERY FOUR BULLETS, SINCE ONLY ONE OF FOUR WAS A TRACER. HOW ANYONE REACHED THE GROUND WITHOUT BEING HIT IS DIFFICULT TO UNDERSTAND.

AFTER WE LANDED, OTHER PLANES CAME OVER IN GOOD FORMATION AND WITH THEIR PARATROOPERS BEING OUT OF REACH OF THE GROUND FIRE. THESE PLANES WERE FROM THE 82ND AIRBORNE AND ST.MERE EGLISE WAS THEIR OBJECTIVE. WE FEW OF THE 101ST AIRBORNE HAD BEAT THEM TO THEIR TARGET. IN FACT THE FIRST MAN I MET UP WITH AFTER I GOT OUT OF MY HARNESS WAS FROM THE 82ND.

I DID NOT KNOW WHERE I WAS FOR SOMETIME SO I FOUGHT THE WAR WHERE NEEDED. WE HAD LANDED ABOUT SIX TO SEVEN HOURS PRIOR TO THE BEACH LANDING SO AS I FOUGHT MY WAY TOWARD THE BEACH, I WAS VERY CLOSE TO THE AREA WHERE THE BOMBS HIT THAT WERE DROPPED BY HUNDREDS OF OUR PLANES. AS THE BOMBS CAME DOWN, THEY LOOKED LIKE A SOLID SHIELD, THEY WERE SO THICK. THE GROUND I WAS ON FELT LIKE AN EARTHQUAKE.

WE FOUGHT WHERE NEEDED FOR SEVERAL DAYS, THEN ON JUNE 12, 1944, WE RECEIVED ORDERS FROM 101ST AIRBORNE HEADQUARTERS FOR THE 506TH REGIMENT TO CAPTURE THE TOWN OF CARENTAN. THE TOWN WAS ABOUT SIX MILES FROM WHERE WE WERE AT THAT TIME. IT WAS AN IMPORTANT HIGHWAY AND RAILWAY LOCATION AND VERY HEAVYLY FORTIFIED BY GERMAN INFANTRY AND HEAVY ARMOUR.

I DO NOT KNOW ABOUT THE REST OF THE 506TH REGIMENT BUT TO THE BEST OF MY MEMORY. THE 2ND BATTALION, CONSISTING OF COMPANIES D-(DOG)--E-(EASY)--(LATER KNOWN AS "BAND OF BROTHERS", AND F=FOX "MY COMPANY"., STARTED OUR ATTACK EARLY MORNING JUNE 13TH, 1944.

WE FIRST FORMED A LINE BEHIND A HEDGEROW, WITH A PLAN TO ADVANCE TOWARD CARENTAN, BY FIGHTING FROM ONE HEDGEROW TO THE NEXT. AS I RECALL, DOG COMPANY WAS ON THE LEFT FLANK. WE, FOX COMPANY, WERE SPLIT IN HALF. WE ONLY HAD ABOUT 70 TO 80 MEN BECAUSE ABOUT 50 HAD BEEN KILLED OR WOUNDED.

A GRAVEL ROAD THAT RAN TOWARD CARENTAN WAS IN THE CENTER OF WHERE WE WERE HEADED, SO THIS IS THE REASON FOR FOX COMPANY TO BE SPLIT INTO TWO SECTIONS. EASY COMPANY WAS ON OUR RIGHT.

I WAS WITH THE PART OF FOX COMPANY THAT WAS ON THE RIGHT SIDE OF THE ROAD. A DRAINAGE DITCH RAN ALONG EACH SIDE OF THE ROAD.

JUST AS WE STARTED TO MAKE OUR ADVANCE OVER THE FIRST HEDGEROW, A TIGER TANK APPEARED ON THE ROAD JUST OVER A SMALL HILL. THEN IN A FEW MINUTES. I SAW GERMAN SOLDIERS CRAWLING ALONG THE DITCHES NEAR THE TANK. I LAID MY M-1 RIFLE ON TOP OF THE HEDGEROW IN FRONT OF ME AND PICKED OUT THE GERMAN SOLDIER I COULD SEE BEST. I FIRED. I WAS SURE I HAD HIT HIM, THE WAY HE SLUMPED. A FEW SE CONDS AFTER I FIRED, THE TIGER TANK FIRED AN 88 SHELL THAT HIT THE EARTHBEARM JUST IN FRONT OF WHERE I AND THREE OTHER TROOPERS WERE LOCATED. IT KNOCKED ALL FOUR OF US OUT. THE OTHER TROOPERS THOUGHT THE FOUR OF US WERE DEAD. AS IT TURNED OUT, TWO WERE DEAD AND I AND ONE OTHER WERE JUST KNOCKED OUT FROM CONCUSION.

AFTER I CAME TO, THE TIGER TANK BACKED UP AND THE GERMANS ALONG THE DITCHES PULLED BACK ALSO. AFTER THEY GOT OUT OF SIGHT, WE MOVED FORWARD. I WENT DOWN TO THE SPOT WHERE I FELT CERTAIN I HAD HIT ONE OF THEM. I WAS RIGHT. IT WAS A GERMAN OFFICER AND HE WAS CARRING A LUGER. I TOOK THE LUGER AND AMMUNITION. I CARRIED AND USED THE LUGER THE REST OF THE WAR.

I USED IT WHEN WE JUMPED IN HOLLAND, WHEN SURROUNDED IN BASTOGNE, FIGHTING ALONG SOUTHERN FRANCE AND GERMANY. I HID IT IN MY CLOTHING WHEN I CAME HOME AND STILL HAVE IT TODAY. EACH TIME I LOOK AT THAT WEAPON, IT BRINGS BACK A LOT OF MEMORIES.

MY OLDEST SON, RONALD WAYNE ALLEY, HAS BEEN RESEARCHING THE HISTORY OF GERMAN WEAPONS AND HAS LEARNED A GREAT DEAL ABOUT THE LUGER I HAVE. HE HAS BEEN VERY HELPFUL. HE TELLS ME THAT MINE IS FROM 1916

THOMAS HERMAN ALLEY,JR.
CO. F 2ND BATTALION
506TH PARACHUTE INFANTRY REGIMENT
101ST AIRBORNE DIVISION

I attached a few photos of his 88th birthday celebration. One of him thumbing your book which he read cover to cover and said to give you his thanks. Another of him "clowning" around in his modesty jacket with his Luger that we had serviced, cleaned, and gently "restored" not refinished. It fires and feeds beautifully. The other photo has me in it. Contributed by Ron Alley

I was walking through a gun show at the San Francisco Cow Palace years ago when I saw this SKS. I thought it was overpriced so passed it by. Later I came back and looked at it more closely. While it was still pretty pricey, I noticed that the dealer was the same person listed on the capture papers. We struck up a conversation as I was very curious as to why he was selling the rifle, and asked to hear the story of its capture. At the time of the capture the dealer was a 1st Lieutenant with 2/4 Arty, and was the FO to Alpha Company, 5th Battalion 60th Infantry. He stated that they were operating in the area known as the "Plain of Reeds" in the Mekong Delta. He stated:

"My RTO and I were walking behind the company command group, providing rear security and blowing mud bunkers with grenades. As the command group passed by a line of concealed bunkers a North Vietnamese soldier emerged from a bunker to ambush the command group. I ran left to get a clear shot at him and as I did he swung toward me. We fired simultaneously, I was hit in the right flank and he was killed. I was evacuated to the 3rd Surgical Hospital in Saigon, and while I was recuperating my company commander and two of the platoon leaders came to visit. The commander handed me the SKS saying "I thought you might like to have the rifle that shot you."

Tom Prall—Grandpa's Japanese Helmet

My grandpa's name is Ivan Prall. He was an Army Air Corps photographer and was part of the 1st Provisional Station Complement Squadron.

Part of Ivan's job was to assist in the set up the 1st airfield on Iwo Jima, and so was stationed in the vicinity of that area. At this time, Ivan could not assist with the setting up of the airfield due to the proximity of the Japanese and their shelling, so he and two other men decided to walk toward the battle front. Ivan stated that while he and the two soldiers were walking to the front, a Japanese soldier from behind a rock had trained his rifle on them. A U.S. marine, who was on higher ground, saw the Japanese soldier and shot at him, knocking the Japanese soldier's helmet off of his head. After the helmet was shot off his head, the Japanese soldier fled back into a cave that was close to where he was attempting to shoot at Ivan and the two other soldiers. Ivan spotted the fallen helmet and picked it up, continuing to the front. Ivan noted that there were several caves along the route that he had taken to the front.

Photography by Eric Lutzens.

These are some interesting shots. They are of the August 19th 1945 surrender of the Japanese military forces on Luzon in the Philippines. This was two weeks before the formal ceremony in Tokyo Bay. The picture and the August 19th surrender were kept secret until the 1990s (I think). Dad always told me these photos were smuggled out. I have found verification of the date this event on the internet.